Ernest Ingersoll

The Crest of the Continent

A record of a summer's ramble in the Rocky Mountains and beyond

Ernest Ingersoll

The Crest of the Continent
A record of a summer's ramble in the Rocky Mountains and beyond

ISBN/EAN: 9783337316112

Printed in Europe, USA, Canada, Australia, Japan

Cover: Foto ©Andreas Hilbeck / pixelio.de

More available books at **www.hansebooks.com**

GARFIELD PEAK.

THE
Crest of the Continent:

A RECORD OF

A SUMMER'S RAMBLE IN THE ROCKY MOUNTAINS AND BEYOND.

By Ernest Ingersoll.

> "We climbed the rock-built breasts of earth!
> We saw the snowy mountains rolled
> Like mighty billows; saw the birth
> Of sudden dawn; beheld the gold
> Of awful sunsets; saw the face
> Of God, and named it boundless space."

CHICAGO:
R. R. DONNELLEY & SONS, PUBLISHERS.
1885

TO
THE PEOPLE OF COLORADO,
SAGACIOUS IN PERCEIVING, DILIGENT IN DEVELOPING,
AND WISE IN ENJOYING
THE
RESOURCES AND ATTRACTIONS OF THE ROCKY MOUNTAINS,
THIS BOOK IS DEDICATED
WITH
THE HOMAGE OF
THE AUTHOR.

PREFACE.

PROBABLY nothing in this artificial world is more deceptive than absolute candor. Hence, though the ensuing text may lack nothing in straightforwardness of assertion, and seem impossible to misunderstand, it may be worth while to say distinctly, here at the start, that it is all true. We actually *did* make such an excursion, in such cars, and with such equipments, as I have described; and we would like to do it again.

It was wild and rough in many respects. Re-arranging the trip, luxuries might be added, and certain inconveniences avoided; but I doubt whether, in so doing, we should greatly increase the pleasure or the profit.

"No man should desire a soft life," wrote King Ælfred the Great. Roughing it, within reasonable grounds, is the marrow of this sort of recreation. What a pungent and wholesome savor to the healthy taste there is in the very phrase! The zest with which one goes about an expedition of any kind in the Rocky Mountains is phenomenal in itself; I despair of making it credited or comprehended by inexperienced lowlanders. We are told that the joys of Paradise will not only actually be greater than earthly pleasures, but that they will be further magnified by our increased spiritual sensitiveness to the "good times" of heaven. Well, in the same way, the senses are so quickened by the clear, vivifying climate of the western uplands in summer, that an experience is tenfold more pleasurable there than it could become in the Mississippi valley. I elsewhere have had something to say about this exhilaration of body and soul in the high Rockies, which you will perhaps pardon me for repeating briefly, for it was written honestly, long ago, and outside of the present connection.

"At sunrise breakfast is over, the mules and everybody else have been good-natured and you feel the glory of mere existence as you vault into the saddle and break into a gallop. Not that this or that particular day is so different from other pleasant mornings, but all that we call *the weather* is constituted in the most perfect proportions. The air is 'nimble and sweet,' and you ride gayly across meadows, through sunny woods of pine and aspen, and between granite knolls that are piled up in the most noble and romantic proportions. . .

"Sometimes it seems, when camp is reached, that one hardly has strength to make another move; but after dinner one finds himself able and willing to do a great deal. . .

"One's sleep in the crisp air, after the fatigues of the day, is sound and serene. . . You awake at daylight a little chilly, re-adjust your blankets, and want again to sleep. The sun may pour forth from the 'golden window of the east' and flood the world with limpid light; the stars may pale and the jet of the midnight sky be diluted to that pale and perfect morning-blue into which you gaze to unmeasured depths; the air may become a pervading Champagne, dry and delicate, every draught of which tingles the lungs and spurs the blood along the veins with joyous speed; the landscape may woo the eyes with airy undulations of prairie or snow-pointed pinnacles lifted sharply against the azure—yet sleep chains you. That very quality of the atmosphere which contributes to all this beauty and makes it so delicious to be awake, makes it equally blessed to slumber. Lying there in the open air, breathing the pure elixir of the untainted mountains, you come to think even the confinement of a flapping tent oppressive, and the ventilation of a sheltering spruce-bough bad."

That was written out of a sincere enthusiasm, which made as naught a whole season's hardship and work, before there was hardly a wagon-road, much less a railway, beyond the front range.

This exordium, my friendly reader, is all to show to you: That we went to the Rockies and beyond them, as we say we did; that we knew what we were after, and found the apples of these Hesperides not dust and ashes but veritable golden fruit; and, finally, that you may be persuaded to test for yourself this natural and lasting enjoyment.

The grand and alluring mountains are still there,—everlasting hills, unchangeable refuges from weariness, anxiety and strife! The railway grows wider and permits a longer and even more varied journey than was ours. Cars can be fitted up as we fitted ours, or in a way as much better as you like. Year by year the facilities for wayside comforts and short branch-excursions are multiplied, with the increase of population and culture.

If you are unable, or do not choose, to undertake all this preparation, I still urge upon you the pleasure and utility of going to the Rocky Mountains, travelling into their mighty heart in comfortable and luxurious public conveyances. Nowhere will a holiday count for more in rest, and in food for subsequent thought and recollection.

CONTENTS.

I — AT THE BASE OF THE ROCKIES.

First Impressions of the Mountains. A Problem, and its Solution. Denver—Descriptive and Historical. The Resources which Assure its Future. Some General Information concerning the Mining, Stock Raising and Agricultural Interests of Colorado. - - - - - - - - - 13

II — ALONG THE FOOTHILLS.

The Expedition Moves. Its Personnel. The Romantic Attractions of the Divide. Light on Monument Park. Colorado Springs, a City of Homes, of Morality and Culture. Its Pleasant Environs: Glen Eyrie, Blair Athol, Austin's Glen, the Cheyenne Cañons. - - - - - - - - 26

III — A MOUNTAIN SPA.

Manitou, and the Mineral Springs. The Ascent of Pike's Peak; bronchos and blue noses. Ute Pass, and Rainbow Falls. The Garden of the Gods. Manitou Park. Williams' Cañon, and the Cave of the Winds. An Indian Legend. - - - 36

IV — PUEBLO AND ITS FURNACES.

The Largest Smelter in the World. The Colorado Coal and Iron Company. Pueblo's Claims as a Trade Center, and its Tributary Railway System. A Chapter of Facts and Figures in support of the New Pittsburgh. - - - - 51

V — OVER THE SANGRE DE CRISTO.

Up and down Veta Mountain, with some Extracts from a letter. Veta Pass, and the Muleshoe Curve. Spanish Peaks. Beautiful Scenery, and Famous Railroading. A general outline of the Rocky Mountain Ranges. - - - - 60

VI — SAN LUIS PARK.

A Fertile and Well-watered Valley. The Method of Irrigation. Sierra Blanca. A Digression to describe the Home on Wheels. Alamosa, Antonito and Conejos. Cattle, Sheep and Agriculture in the largest Mountain Park. - - 71

VII — THE INVASION OF NEW MEXICO.

Barranca, among the Sunflowers. An Excursion to Ojo Caliente, and Description of the Hot Springs. Pre-historic Relics—a Rich Field for the Archæologist. Señor vs. Burro. An Ancient Church, with its Sacred Images. - - - 81

VIII — El Mexicano y Puebloano.

Comanche Cañon and Embudo. The Penitentes. The Rio Grande Valley; Alcalde, Chamita and Espanola. New Mexican Life, Homes and Industries. The Indian Pueblos, and their Strange History. Architecture, Pottery, and Threshing. 92

IX — Santa Fe and the Sacred Valley.

Santa Fe, the Oldest City in the United States. Fact and Tradition. San Fernandez de Taos—the Home of Kit Carson. Pueblo de Taos Birthplace of Montezuma, and Typical and Well-Preserved. The Festival of St. Geronimo. Exit Amos. - 106

X — Toltec Gorge.

Heading for the San Juan Country. From Mesa to Mountain Top. The Curl of the Whiplash. Above the silvery Los Pinos. Phantom Curve. A Startling Peep from Toltec Tunnel. Eva Cliff. "In Memoriam." - - - - 115

XI — Along the Southern Border.

The Piños-Chama Summit. Trout and Game. The Groves of Chama. Mexican Rural Life at Tierra Amarilla. The Iron Trail. Rio San Juan and its Tributaries. Pagosa Springs. Apache Visitors. The Southern Utes. Durango. - 120

XII — The Queen of the Cañons.

Geology of the Sierra San Juan. The Attractions of Trimble Springs. Beauty and Fertility of the Animas Valley. The Cañon of the River of Lost Souls. Engineering under difficulties. The Needles, and Garfield Peak. - - 129

XIII — Silver San Juan.

Geological Resume. Scraps of History. Snow-shoes and Avalanches. The Mining Camps of Animas Forks, Mineral Point, Eureka and Howardville. Early Days in Baker's Park. Poughkeepsie, Picayune and Cunningham Gulches. The Hanging. - - - - - - - 136

XIV — Beyond the Ranges.

Ophir, Rico, and the La Plata Mountains. Everything triangular. Mixed Mineralogy, Real bits of Beauty. "When I sell my Mine." An Unbiased Opinion. Placer vs. Fissure Vein Mining. - - - - - 149

XV — The Antiquities of the Rio San Juan.

Rugged Trails. Searching for Antiquities. The Discovery. Habitations of a Lost Race. Prehistoric Architecture, "Temple or Refrigerator." "Ruins, Ancient beyond all Knowing." Guesses and Traditions. Some Appropriate Verses. 156

XVI — On the Upper Rio Grande.

Good-bye and Welcome. Del Norte and the Gold Summit. Among the River Ranches. Wagon Wheel Springs. Healing Power of the Waters. The Gap and its History. A Day's Trout Fishing. - - - - - 166

XVII — El Moro and Cañon City.

A Great Natural Fortress. Down in a Coal Mine. The Coke Ovens. Huerfano Park and its Coal. Cañon City Historically. Coal Measures. Resources of the Foot-hills. - - - - - - 177

CONTENTS.

XVIII — IN THE WET MOUNTAIN VALLEY.

Grape Creek Cañon. The Dome of the Temple. Wet Mountain Valley. The Legend of Rosita. Hardscrabble District. Silver Cliff and its Strange Mine. The Foothills of the Sierra Mojada. Geological Theories. - - - 185

XIX — THE ROYAL GORGE.

The Grand Cañon of the Arkansas. Its Culminating Chasm the Royal Gorge. Beetling Cliffs and Narrow Waters. Running the Gauntlet. Wonders of Plutonic Force. A Story of the Cañon. - - - 193

XX — THE ARKANSAS VALLEY.

Entering Brown's Cañon. The Iron Mines of Calumet. Salida. Farming on the Arkansas. Buena Vista. Granate and its Gold Placers,—Twin Lakes. Malta and its Charcoal Burners. A Burned-out Gulch. - - - 201

XXI — CAMP OF THE CARBONATES.

California Gulch. How Boughtown was Built. Some Lively Scenes. Discovery of Carbonates. The Rush of 1878. The Founding of Leadville. A Happy Grave Digger. Practice and Theory of Mining. Reducing the Ores. - - 209

XXII — ACROSS THE TENNESSEE AND FREMONT'S PASS.

Hay Meadows on the Upper Arkansas. Climbing Tennessee Pass. Mount of the Holy Cross. Red Cliff. Ore in Battle Mountain. Through Eagle River Cañon. The Artist's Elysium. Two Miles in the Air. On the Blue. - - 222

XXIII — FROM PONCHO SPRINGS TO VILLA GROVE.

In Hot Water. A Pretty Village and Fine Outlook. Pluto's Reservoirs. The Madame's Letter. Poncho Pass. The Sangre de Cristo Again. Villa Grove. Silver and Iron. - - - 225

XXIV — THROUGH MARSHALL PASS.

The Unknown Gunnison. A Wonder of Progress. Climbing the Mountains in a Parlor Car. Four Hours of Scenic Delight. Culmination of Man's Skill. On the Crest of the Continent. The Mysterious Descent. - - - 243

XXV — GUNNISON AND CRESTED BUTTE.

Tomichi Valley. Gunnison from Oregon to St. Louis. Captain Gunnison's Discoveries. A Discussion with Chief Ouray. A Beautiful Landscape. Crested Butte. Anthracite in the Rockies. - - - 250

XXVI — A TRIP TO LAKE CITY.

Lake City. A Picture from Nature. A Hard Pillow. The Mining Interests. Alpine Grandeur of the Scenery. The Home of the Bear and the Elk. Game, Game, Game. - - - 262

XXVII — IMPRESSIONS OF THE BLACK CAÑON.

The Observation Car. Gunnison River. Trout Fishing Again. The Rock Cleft in Twain. A Beautiful Cataract. A Mighty Needle. The Cañon Black yet Sunny. Impressions of the Cañon. Majestic Forms and Splendid Colors. - - 266

XXVIII — THE UNCOMPAHGRE VALLEY.

Cline's Ranch. Montrose. The Madame and Chum Respectfully Decline. The Trip to Ouray. The Military Post. Chief Ouray's Widow. The Road on the Bluff. Hot Springs. Brilliant Stars. - 273

XXIX — OURAY AND RED MOUNTAIN.

A Pretty Mountain Town. Trials of the Prospectors. A Tradition. From Silverton to Ouray by Wagon. Enchanting Gorges and Alluring Peaks. The Yankee Girl. A Cave of Carbonates. Vermillion Cliffs. Dallas Station. - . 278

XXX — MONTROSE AND DELTA.

Playing Billiards. Caught in the Act. A Well-Watered District. Coal and Cattle. A Fruit Garden. A Big Irrigating Ditch. The Snowy Elk Mountains. A Substantial Track. A Long Bridge. - 290

XXXI — THE GRAND RIVER VALLEY.

An Honest Circular. Grand Junction. Staking Out Ranches. The Recipe for Good Soil. Watering the Valley. Value of Water. Some Big Corn in the Far West. A Land of Plenty. Going West. - 296

XXXII — THE COLORADO CAÑONS.

A Memorable Night-Journey. Skirting the Uncompahgre Plateau. Origin of the Sierra La Sal. Crossing the Green River. Wonders of Erosive Work. An Indian Tradition. The Marvelous Cañons of the Colorado. - . . . 303

XXXIII — CROSSING THE WASATCH.

The Tall Cliffs of Price River and Castle Cañon. Castle Gate. The Summit of the Wasatch. "Indians!" San Pete and Sevier Valleys. "Like Iser Rolling Rapidly." Through the Cañon of the Spanish Fork. Mount Nebo. - . . 312

XXXIV — BY UTAH LAKES.

Rural Scenes Beside Lake Utah. Spanish Fork, Springville, Provo and Nephi. Relics of Indian Wars. Pretty Fruit Sellers. First Sight of Deseret and the Great Salt Lake. Ogden and Its History. - 317

XXXV — SALT LAKE CITY.

Sunday in Salt Lake City. The Tabernacle and the Temple. Early Days in Utah. Shady Trees and Sparkling Brooks. Social Peculiarities of the City. Mining and Mercantile Prosperity. Religious Sects. Schools and Seminaries. . 324

XXXVI — SALT LAKE AND THE WASATCH.

The Ride to Salt Lake. A Salt Water Bath. Keep Your Mouth Shut. The Shore of the Lake. An Exciting Chase. A Trip to Alta. Stone for the Temple. An Exhilarating Ride. - 335

XXXVII — AU REVOIR.

At Last. On Jordan's Banks. Chum's Grandfather. Let Every Injun Carry his Own Skillet. The Parting Toast. Good-Night. - . . . 342

ILLUSTRATIONS.

	PAGE
GARFIELD PEAK	*Frontispiece.*
DENVER	17
DEPOT AT PALMER LAKE	20
PHŒBE'S ARCH	21
MONUMENT PARK	24
IN QUEEN'S CAÑON	28
CHEYENNE FALLS	31
IN NORTH CHEYENNE CAÑON	34
A GLIMPSE OF MANITOU AND PIKE'S PEAK	37
THE MINERAL SPRINGS	40
PIKE'S PEAK TRAIL	45
RAINBOW FALLS	49
GARDEN OF THE GODS	53
ENTRANCE TO CAVE OF THE WINDS	57
ALABASTER HALL	62
VETA PASS	67
CREST OF VETA MOUNTAIN	69
SPANISH PEAKS FROM VETA PASS	75
SANGRE DE CRISTO SUMMITS	78
SIERRA BLANCA	83
OJO CALIENTE	86
EMBUDO, RIO GRANDE VALLEY	89
NEW MEXICAN LIFE	94
A PATRIARCH	98
MAID AND MATRON	99
OLD CHURCH OF SAN JUAN	102
PUEBLO DE TAOS	107
PHANTOM CURVE	112
PHANTOM ROCKS	113
IN MEMORIAM	119
TOLTEC GORGE	125
EVA CLIFF	130
GARFIELD MEMORIAL	131
NEAR THE PIÑOS-CHAMA SUMMIT	136
CHIEFS OF THE SOUTHERN UTES	141
CAÑON OF THE RIO DE LAS ANIMAS	146
ON THE RIVER OF LOST SOULS	152
ANIMAS CAÑON AND THE NEEDLES	157

	PAGE
Silverton and Sultan Mountain	162
Cliff Dwellings	168
Wagon Wheel Gap	173
Up the Rio Grande	178
Grape Creek Cañon	181
Grand Cañon of the Arkansas	186
The Royal Gorge	191
Brown's Cañon	194
Twin Lakes	199
The Old Route to Leadville	202
The Shaft House	204
Bottom of the Shaft	205
Athwart an Incline	206
The Jig Drill	207
Fremont Pass	211
Cascades of the Blue	214
Mount of the Holy Cross	219
Marshall Pass—Eastern Slope	223
Marshall Pass—Western Slope	227
Crested Butte Mountain and Lake	230
Ruby Falls	232
Approach to the Black Cañon	235
Black Cañon of the Gunnison	241
Currecanti Needle, Black Cañon	247
A Ute Council Fire	251
Ouray	255
Gate of Lodore	261
Winnie's Grotto	264
Echo Rock	267
Gunnison's Butte	271
Buttes of the Cross	274
Marble Cañon	279
Grand Cañon of the Colorado	283
Grand Cañon, from To-ro-wasp	287
Exploring the Walls	292
Castle Gate	297
In Spanish Fork Cañon	300
Tramway in Little Cottonwood Cañon	305
Salt Lake City	311
Mormon Temple, Tabernacle and Assembly Hall	325
Great Salt Lake	331

I

AT THE BASE OF THE ROCKIES.

OLD WOODCOCK says that if Providence had not made him a justice of the peace, he'd have been a vagabond himself. No such kind interference prevailed in my case. I was a vagabond from my cradle. I never could be sent to school alone like other children —they always had to see me there safe, and fetch me back again. The rambling bump monopolized my whole head. I am sure my godfather must have been the Wandering Jew or a king's messenger. Here I am again, *en route*, and sorely puzzled to know whither.—
THE LOITERINGS OF ARTHUR O'LEARY.

HERE are the Rocky Mountains!' I strained my eyes in the direction of his finger, but for a minute could see nothing. Presently sight became adjusted to a new focus, and out against a bright sky dawned slowly the undefined shimmering trace of something a little bluer. Still it seemed nothing tangible. It might have passed for a vapor effect of the horizon, had not the driver called it otherwise. Another minute and it took slightly more certain shape. It cannot be described by any Eastern analogy; no other far mountain view that I ever saw is at all like it. If you have seen those sea-side albums which ladies fill with algæ during their summer holiday, and in those albums have been startled, on turning over a page suddenly, to see an exquisite marine ghost appear, almost evanescent in its faint azure, but still a literal existence, which had been called up from the deeps, and laid to rest with infinite delicacy and difficulty,—then you will form some conception of the first view of the Rocky Mountains. It is impossible to imagine them built of earth, rock, anything terrestrial; to fancy them cloven by horrible chasms, or shaggy with giant woods. They are made out of the air and the sunshine which show them. Nature has dipped her pencil in the faintest solution of ultramarine, and drawn it once across the Western sky with a hand tender as Love's. Then when sight becomes still better adjusted, you find the most delicate division taking place in this pale blot of beauty, near its upper edge. It is rimmed with a mere thread of opaline and crystalline light. For a moment it sways before you and is confused. But your eagerness grows steadier, you see plainer and know that you are looking on the everlasting snow, the ice that never melts. As the entire fact in all its meaning possesses you completely, you feel a sensation which is as new to your life as it is impossible of repetition. I confess (I should be ashamed not to) that my first

view of the Rocky Mountains had no way of expressing itself save in tears. To see what they looked, and to know what they were, was like a sudden revelation of the truth that the spiritual is the only real and substantial; that the eternal things of the universe are they which, afar off, seem dim and faint."

There are the Rocky Mountains! Ludlow saw them after days of rough riding in a dusty stage-coach. Our plains journey had been a matter of a few hours only, and in the luxurious ease of a Pullman sleeping car; but *our* hearts, too, were stirred, and we eagerly watched them rise higher and higher, and perfect their ranks, as we threaded the bluffs into Pueblo. Then there they were again, all the way up to Denver; and when we arose in the morning and glanced out of the hotel window, the first objects our glad eyes rested on were the snow-tipped peaks filling the horizon.

Thither *Madame ma femme* and I proposed to ourselves to go for an early autumn ramble, gathering such friends and accomplices as presented themselves. But how? That required some study. There were no end of ways. We were given advice enough to make a substantial appendix to the present volume, though I suspect that it would be as useless to print it for you as it was to talk it to us. We could walk. We could tramp, with burros to carry our luggage, and with or without other burros to carry ourselves. We could form an alliance, offensive and defensive, with a number of pack mules. We could hire an ambulance sort of wagon, with bedroom and kitchen and all the other attachments. We could go by railway to certain points, and there diverge. Or, as one sober youth suggested, we needn't go at all. But it remained for us to solve the problem after all. As generally happens in this life of ours, the fellow who gets on owes it to his own momentum, for the most part. It came upon us quite by inspiration. We jumped to the conclusion; which, as the Madame truly observed, is not altogether wrong if only you look before you leap. That is a good specimen of feminine logic in general, and the Madame's in particular.

But what was the inspiration — the conclusion — the decision? You are all impatience to know it, of course. It was this:

Charter a train!

Recovering our senses after this startling generalization, particulars came in order. Spreading out the crisp and squarely-folded map of Colorado, we began to study it with novel interest, and very quickly discovered that if our brilliant inspiration was really to be executed, we must confine ourselves to the narrow-gauge lines. Tracing these with one prong of a hair pin, it was apparent that they ran almost everywhere in the mountainous parts of the State, and where they did not go now they were projected for speedy completion. Closer inspection, as to the names of the lines, discovered that nearly all of this wide-branching system bore the mystical letters D. & R. G., which evidently enough (after you had learned it) stood for—

"Why, Denver and Ryo Grand, of course," exclaims the Madame, contemptuous of any one who didn't know *that*.

"Not by a long shot!" I reply triumphantly, "Denver and Reco Grandy is the name of the railway—Mexican words."

"Oh, indeed!" is what I *hear;* a very lofty nose, naturally a trifle uppish, is what I *see*.

Deciding that our best plan is to take counsel with the officers of the Denver and Rio Grande railway, we go immediately to interview Mr. Hooper, the General Passenger Agent, among whose many duties is that of receiving, counseling, and arranging itineraries for all sorts of pilgrims. An hour's discussion perfected our arrangements, and set the workmen at the shops busy in preparing the cars for our migratory residence.

The realization that our scheme, which up to this point had seemed akin to a wild dream, was now rapidly growing into a promising reality, did not diminish our enthusiasm. Indeed we experienced an exhilaration which was quite phenomenal. Was it the very light wine we partook at luncheon? Perish the suspicion! Possibly it was the popularly asserted effect of the rarefied atmosphere. But kinder to our self-esteem than either of these was the thought that our approaching journey had something to do with our elevation, and we accepted it as an explanation.

But we had yet a few days to spare, and we could employ them profitably in looking over this Denver, the marvelous city of the plains. We studied it first from Capitol Hill, as our artist has done, though his picture, so excellently reproduced, can convey but the shadow of the substance. Then we nearly encompassed the town, going southward on Broadway until we had passed Cherry Creek, and *détouring* across Platte River to the westward and northward, on the high plateau which stretches away to the foothills. The city lies at an altitude of 5,197 feet, near the western border of the plains, and within twelve miles of the mountains,—the Colorado or front range of which may be seen for an extent of over two hundred miles. In the north, Long's Peak rears its majestic proportions against the azure sky. Westward, Mounts Rosalie and Evans rise grandly above the other summits of the snowy range, and Gray's and James' Peaks peer from among their gigantic brethren; while historic Pike's Peak, the mighty landmark that guided the gold-hunters of '59, plainly shows its white crest eighty miles to the south. The great plains stretch out for hundreds of miles to the north, east and south. Near the smelting works at Argo, we retrace our way and re-enter the city.

It is the metropolis of the Rocky Mountains, and a stroll through these scores of solid blocks of salesrooms and factories exhibits at once the fact that it is as the commercial center of the mountainous interior that Denver thrives, and congratulates herself upon the promise of a continually prosperous future. She long ago safely passed that crisis which has proved fatal to so many incipient Western cities. Every year proves anew the wisdom and foresight of her founders; and I think her assertion that she is to be the largest city between Chicago and San Francisco is likely to be realized. Most of her leading business men came here at the beginning, but their energies were hampered when every article had to be hauled six hundred miles across the plains by teams. It frequently used to happen that merchants would sell their goods completely out, put up their shutters and go a-fishing for weeks, before the new semi-yearly supplies arrived. Everybody therefore looked forward, with good reason, to railway communication as the beginning of a new era of prosperity, and watched with keen interest the approach of the Union Pacific lines from Omaha and Kansas City. These were completed, by the northern routes, in 1869 and 1870; and a few months later the enterprising Atchison, Topeka and Santa Fe sent its tracks and trains through to the mountains, and then came the Burlington route, a most welcome acquisition, adding another link to the transcontinental chain, which now binds the East to the West, the Atlantic to the Pacific coast. At Pueblo, the Denver and Rio Grande, meeting the Atchison, Topeka and Santa Fe, was already prepared to make this new route available to Denver and much of Colorado, and adopting a liberal policy, at once exerted an immense influence upon the speedy development and prosperity of the State.

Thus, in a year or two, the young city found itself removed from total isolation to a central position on various railways, east and west, and to its mill came the varied grist of a circle hundreds of miles in radius. Now blossomed the booming season of business which sagacious eyes had foreseen. The town had less than four thousand inhabitants in 1870. A year from that time her population was nearly fifteen thousand, and her tax-valuation had increased from three to ten millions of dollars. It was a time of happy investment, of incessant building and improvement, and of grand speculation. Mines flourished, crops were abundant, cattle and sheep grazed in a hundred valleys hitherto tenanted by antelope alone, and everybody had plenty of money. Then came a shadow of storm in the East, and the sound of the thunder-clap of 1873 was heard in Denver, if the bolt of the panic was not felt. The banks suddenly became cautious in loans; speculators declined to buy, and sold at a sacrifice. Merchants found that trade was dull, and ranchmen got less for their products. It was a "set-back" to Denver, and two years of stagnation followed. But she only dug the more money out of the ground to fill her depleted pockets, and survived the "hard times" with far less sacrifice of fortune and pride than did most of the Eastern cities. None of her banks went under, nor even certified

DENVER.

a check, and most of her business houses weathered the storm. The unhealthy reign of speculation was effectually checked, and business was placed upon a compact and solid foundation. Then came 1875 and 1876, which were "grasshopper years," when no crops of consequence were raised throughout the State, and a large amount of money was sent East to pay for flour and grain. This was a particularly hard blow, but the bountiful harvest of 1877 compensated, and the export of beeves and sheep, with their wool, hides and tallow, was the largest ever made up to that time.

The issue of this successful year with miner, farmer and stock-ranger, yielding them more than $15,000,000, a large proportion of which was an addition to the intrinsic wealth of the world, had an almost magical effect upon the city. Commerce revived, a buoyant feeling prevailed among all classes, and merchants enjoyed a remunerative trade. Money was "easy," rents advanced, and the real estate business assumed a healthier tone. Generous patronage of the productive industries throughout the whole State was made visible in the quickened trade of the city, which rendered the year an important one in the history of Denver's progress. So, out of the barrenness of the cactus-plain, and through this turbulent history, has arisen a cultivated and beautiful city of 75,000 people, which is truly a metropolis. Her streets are broad, straight, and everywhere well shaded with lines of cotton-woods and maples, abundant in foliage and of graceful shape. On each side of every street flows a constant stream of water, often as clear and cool as a mountain brook. The source is a dozen miles southward, whence the water is conducted in an open channel. There are said to be over 260 miles of these irrigating ditches or gutters, and 250,000 shade-trees.

For many blocks in the southern and western quarter of the town, —from Fourteenth to Thirtieth streets, and from Arapahoe to Broadway and the new suburbs beyond—you will see only elegant and comfortable houses. Homes succeed one another in endlessly varying styles of architecture, and vie in attractiveness, each surrounded by lawns and gardens abounding in flowers. All look new and ornate, while some of the dwellings of wealthy citizens are palatial in size and furnishing, and with porches well occupied during the long, cool twilight characteristic of the summer evening in this climate, giving a very attractive air of opulence and ease. Even the stranger may share in the general enjoyment, for never was there a city with so many and such pleasant hotels, the largest of which, the Windsor and the St. James, are worthy of Broadway or Chestnut street.

The power which has wrought all this change in a short score of years, truly making the desert to bloom, is water; or, more correctly, that is the great instrument used, for the *power* is the will and pride of the intelligent men and women who form the leading portion of the citizens. Water is pumped from the Platte, by the Holly system, and forced over the city with such power that in case of fire no steam-engine is necessary to send a strong stream through the hose. The keeping of a turf and garden, after it is once begun, is merely a matter of watering. The garden is kept moist mainly by flooding from the irrigating ditch in the street or alley, but the turf of the lawn and the shrubbery owe their greenness to almost incessant sprinkling by the hand-hose. Fountains are placed in nearly every yard. After dinner (for Denver dines at five o'clock as a rule), the father of the house lights his cigar and turns hose-man for an hour, while he chats with friends; or the small boys bribe each other to let them lay the dust in the street, to the imminent peril of passers-by; and young ladies escape the too engrossing attention of complimentary admirers by busily sprinkling heliotrope and mignonette, hinting at a possible different use of the weapon if admiration becomes too ardent. The swish and gurgle and sparkle of water are always present, and always must be; for so Denver defies the desert and dissipates the dreaded dust.

Their climate is one of the things Denverites boast of. That the air is pure and invigorating is to be expected at a point right out on a plateau a mile above sea-level, with a range of snow-burdened mountains within sight. From the beginning to the end of warm weather it rarely rains, except occasional thunder and hail storms in July and August. September witnesses a few storms, succeeded by cool, charming weather, when the haze and smoke is filtered from the bracing air, and the landscape robes itself in its most enchanting hues. The coldest weather occurs after New Year's Day, and lasts sometimes until April. Then come the May storms and floods, followed by a charming summer. The barometer holds itself pretty steady throughout the year. There is a vast quantity of electricity in the air, and the displays of lightning are magnificent and occasionally destructive. Sunshine is very abundant. One can by no means judge from the brightest day in New York of the wonderful glow sunlight has here. During 1884 there were 205 clear days, 126 fair, and 34 cloudy, the sun being totally obscured on only 18 days; and yet this record is more unfavorable than the average for a number of years. Summer heat often reaches a hundred in the shade at midday; but with sunset comes coolness, and the nights allow refreshing sleep. In winter the mercury sometimes sinks twenty degrees below zero; but one does not feel this severity as much as he would a far less degree of cold in the damp, raw climate of the coast. Snow is frequent, but rarely plentiful enough for sleighing.

Denver is built not only with the capital of her own citizens, but constructed of materials close at hand. Very substantial bricks, kilned in the suburbs, are the favorite material. Then there is a pinkish trachyte, almost as light as pumice, and ringing under a blow with a metallic clink, that is largely employed in trimmings. Sandstone, marble and limestone are abundant enough for all needs. Coarse lumber is supplied by the high pine forests, but all the hard wood and fine lumber is brought from the East. The fuel of the city was formerly wholly lignite coal, which comes from the foothills; but the extension of the railway to Cañon City, El Moro and the Gunnison, have made the harder and less sulphurous coals accessible and cheap.

And while she has been looking well after the material attractions, Denver has kept pace with the progress of the times in modern advantages. She is very proud of her school-buildings, constructed and managed upon the most improved plans; of her fine churches, of her State and county offices, her seminaries of higher learning, and of her natural history and historical association. Her Grand Opera House is the most elegant on the continent, her business blocks are extensive and costly, and the Union Depot ranks with the best of similar structures. Gas was introduced several years ago, and the system, which now includes nearly all sections of the city, is being constantly improved and extended. The Brush electric light has been in very general use for nearly three years, and the Edison incandescent lamps are now being

DEPOT AT PALMER LAKE.

employed. The telephone is found in hundreds of business places and residences, the exchange at the close of last year numbering 709 subscribers. The water supply is distributed through forty miles of mains, the consumption averaging three million gallons per day, exclusive of the contributions of the irrigating ditches and the numerous artesian wells. The steam heating works evaporate one hundred thousand gallons of water daily, delivering the product through three miles of mains and nearly two miles of service pipes; this being the only company out of twenty seven of its nature in the country which has proved a financial success. Street car lines traverse the thoroughfares in all directions, and transport over two million passengers annually. Two district messenger companies are generously patronized. The regular police force consists of some forty-five patrolmen and detectives, aside from the Chief and his assistants; and a distinct organization is the Merchant's Police, numbering twenty men.

PHEBE'S ARCH.

A paid Fire Department is maintained, at an annual expense of $56,000, and the alarm system embraces twenty-six miles of wire and fifty signal boxes. There are published six daily newspapers, one being in German, and a score of weeklies. All are well conducted and prosperous.

A branch of the United States Mint is located here, but is used for assays only, and not for coinage. An appropriation has been made by Congress for a handsome building, the site has been selected, and work is now being pushed forward. The post-office is a source of considerable revenue to the Government. There are six National and two State banks, with a paid in capital of $870,000, and showing a surplus of $754,000 at the close of 1883. The deposits for the year amounted to $8,396,200, and the loans and discounts approximated $4,500,000. The shops of the Denver and Rio Grande railway are doubtless the most extensive in the West, employing over 800 men, and turning out during the year 2 express, 8 mail, 4 combination, 522 box, 303 stock, 25 refrigerator, 197 flat, and 300 coal cars, together with 8 cabooses. In addition they have produced 350 frogs, 200 switch stands, and all the iron work for the bridges on 350 miles of new road. The year's shipments of the Boston and Colorado Smelting Company aggregate, in silver, gold, and copper, $3,907,000; and in the same time the Grant smelter has treated, in silver, lead and gold, $6,348,868. Finally, from the statistics at hand it appears that the volume of Denver's trade for the year referred to, apart from the industries above mentioned, and real estate transactions, has exceeded the snug sum of $58,856,998. In the meantime the taxable valuation of property in Arapahoe county has increased $6,600,000.

These facts establish, beyond the slightest doubt, the truth that Denver stands upon a firm financial basis. This the casual stranger can hardly fail to surmise when he glances at her magnificent buildings, and statistics will confirm the surmise.

Denver society is cosmopolitan. Famous and brilliant persons are constantly appearing from all quarters of the globe. Five hundred people a day, it is said, enter Colorado, and nine-tenths of this multitude pass through Denver. Nowadays, "the tour" of the United States is incomplete if this mountain city is omitted. Thus the registers of her hotels bear many foreign autographs of world-wide reputation. Surprise is often expressed by the critical among these visitors (why, I do not understand) at the totally unexpected degree of intelligence, appreciation of the more refined methods of thought and handiwork, and the knowledge of science, that greet them here. Matters of art and music, particularly, find friends and cultivation among the educated and generous families who have built up society; and there are schools and associations devoted to sustaining the interest in them, just as there are reading circles and literary clubs. And, withal, there is the most charming freedom of acquaintance and intercourse — the polish and good-breeding of rank, delivered from all chill and exclusiveness or regard for "who was your grandfather." Yet this winsome good-fellowship by no means descends to vulgarity, or permits itself to be abused. After all, it is only New York and New England and Ohio, transplanted and considerably enlivened.

Returning to our consideration of Denver's resources, it will readily be seen that she stands as the supply-depot and money-receiver of three great branches of industry and wealth, namely, mining, stock-raising and agriculture.

The first of these is the most important. Many of the richest proprietors live and spend their profits here. Then, too, the machinery which the mining and the reduction of the ores require, and the tools, clothing and provisions of the men, mainly come from here. Long ago ex-Governor Gilpin, worthily one of the most famous of Colorado's representative men, and an enthusiast upon the subject of her virtues and loveliness, prophesied the immense wealth which would continue to be delved from the crevices of her rocky frame, and was called a visionary for his pains; but his prophesies have aggregated more in the fulfillment than they promised in the foretelling, and his "visions" have netted him a most satisfactory fortune. About 75,000 lodes have been discovered in Colorado, and numberless placers. Only a small proportion of these, of course, were worked remuneratively, but the cash yield of the twenty years since the discovery of the precious metals, has averaged over $7,000,000 a year, and has increased from

$200,000 in 1869 to over $26,376,562 in 1883. Not half of this is gold, yet it is only since 1870 that silver has been mined at all in Colorado. These statistics show the total yield of the State in gold and silver thus far to exceed $154,000,000, not to mention tellurium, copper, iron, lead and coal. Surely this alone is sufficient employment of capital and production of original wealth — genuine making of money — to ensure the permanent support of the city.

The second great source of revenue to Denver, is the cattle and sheep of the State. The wonderful worthless-looking buffalo grass, growing in little tufts so scattered that the dust shows itself everywhere between, and turning sere and shriveled before the spring rains are fairly over, has proved one of Colorado's most prolific avenues of wealth. The herds now reported in the State count up 1,461,945 head, and the annual shipments amount to 100,000, at an average of $20 apiece, giving $2,000,000 as the yearly yield. Add the receipts for the sales of hides and tallow, and the home consumption, amounting to about $60,000, and you have a figure not far from $3,500,000 to represent the total annual income from this branch of productive industry. The whole value of the cattle investments in the State is estimated by good judges at $14,000,000, nearly one-fourth of which is the property of citizens of Denver. Yet this sum, great as it is for a pioneer region, represents only two-thirds of Colorado's live stock. Last year about 1,500,000 sheep were sheared, and more capital is being invested in them. Perhaps the total value of sheep ranches is not less than $5,000,000, the annual income from which approaches $1,300,000.

The third large item of prosperity is agriculture, although it advances in the face of much opposition. In 1883 the production of the chief crops was as follows: hay, 266,500 tons; wheat, 1,750,840 bushels; oats, 1,186,534 bushels; corn, 598,975 bushels; barley, 265,180 bushels; rye, 78,030 bushels, and potatoes, 851,000 bushels. Add to this vegetables and small fruits, and the yield of the soil in Colorado is brought to over $9,000,000 in value. Farmers are learning better and better how to produce the very best results by means of scientific irrigation, and the tillage is annually wider.

Nor is this the whole story. Denver is rapidly growing into a manufacturing center. Here are rolling mills, iron foundries, smelters, machine shops, woolen mills, shoe factories, glass works, carriage and harness factories, breweries, and so on through a long list. The flouring mills are very valuable, representing an investment of $950,000, and handling half the wheat crop of Colorado. I have dwelt upon these somewhat prosy statements in order to point out fully what rich resources Denver has behind her, and how it happens that she finds herself, at twenty-three years of age, amazingly strong commercially. Not only a large proportion of the money which gives existence to these enterprises (nearly every householder in the city has a financial interest in one or several mines, stock-ranges or farms), but, as I have intimated, the current supplies that sustain them, are procured in Denver, and a very large percentage of their profits finds its way directly to this focus.

Denver thus becomes to all Colorado what Paris is to France.

MONUMENT PARK.

Through all the enormous area, from Wyoming far into New Mexico, and westward to Utah, she has had no formidable rival until South Pueblo rose to contest the trade of all the southern half of this commercial territory. That she advances with the rapidly thickening population of the State and its increasing needs, is apparent to every one who has noted the gigantic strides with which Denver has grown, and the ease with which she wears her imperial honors. Every extension of the railways, every good crop, every new mineral district developed, every increase of stock ranges, directly and instantly affects the great central mart. This sound business basis being present, the opportunity to pleasantly dispose of the money made is, of course, not long in presenting itself. It thus happens that Denver shows, in a wonderful measure, the amenities of intellectual culture that make life so attractive in the old-established centers of civilization, where selected society, thoughtful study, and the riches of art, have ripened to maturity through long time and under gracious traditions. There is an abundance here, therefore, to please the eye and touch the heart as well as fill the pockets, and year by year the city is becoming more and more a desirable place in which to dwell as well as to do business.

II

ALONG THE FOOTHILLS.

> We've left behind the busy town,
> Its woof and warp of care;
> Our course is down the foothills brown
> To a Southern city fair.
>
> —STANLEY H. RAY.

WHILE we were codifying our impressions of Denver, the workmen at the shops had been busy. We were busy, too, in other than literary ways, and badgered our new acquaintances at the railway offices at all sorts of times and with every manner of want. The butcher and baker were harassed, and jolly old Salomon, the grocer, came in for his share of the nuisance. But it didn't last long, for one afternoon, just three days from the birth of the happy thought, we were in our special train and rushing away to the South.

Not till then did this haphazard crowd — for we had enlisted three gentlemen into our company — inquire seriously whither we were going. What did it matter? We were wild with joy because of going at all. Had we not bed and provender with us? Why could we not go on always? have it said of us when living, *Going, going*, and written over us when dead — *Gone!*

I have mentioned three companions besides the Madame. At least two of the gentlemen you would recognize at once, were I to give you their names. The Artist is famed on both sides of the Atlantic for the masterly productions of his brush. He is a wide traveler and an enthusiast over mountain scenery. The Photographer is likewise a genius, and literally a compendium of scientific knowledge and exploration. Connected for many years with the Geological Surveys of this region, his practical experience renders him an especially valuable coadjutor. The Musician is young in years, with the scroll of fame before him. But he comes of good stock, and faith is strong. And there is still another, our Amos, of sable hue, who has our fortunes, to a large extent, in his keeping, for does he not preside over our commissary? We shall know him better by and by.

Our train consisted of three cars; and when we had passed the great works at Burnham, we resolved ourselves into an investigating committee and started on a tour of inspection. We found our quarters exceedingly well-arranged and comfortable, although in some confusion from the hasty manner in which our loose supplies had been tumbled in. So the committee postponed its report.

"How smoothly we bowl along!" remarked the Musician.

"And in what superb condition are the roadbed and track," added the Photographer.

Yet so gentle and noiseless was the motion that it required the testimony of the speed-indicator to convince us that we were making thirty-five miles per hour. We had passed the huge Exposition building at our left, flitted by the picturesque village of Littleton, with its neat stone depot and white flouring-mills, and were approaching Acequia (which you must pronounce A-sáy-ke-a) along a shelving embankment overlooking the Platte. Away to the west and across the valley, we could discern a yellow band, which the Photographer explained was the new canal under construction by an English company, and which was intended to convey the water of the Platte, from a point far up its cañon, to Denver. The canal or ditch here emerges from the mountains and bears away to the southward for some distance, until it nears Plum Creek, crossing which, by means of an aqueduct, it turns sharply to the northward, and *apparently climbs* the higher table-lands in the direction of Denver. As observed from the car window the anomaly seems indisputable, the deception of course being attributable to the ascending grade of the railway. This is one of several cases in the State which will be pointed out, by old-timers to new-comers, as veritable instances where water runs up hill.

The valleys of Plum Creek and its branches are of good width, and hollowed out of the modern deposits so as to form beautiful and fertile lands, while on each side a terrace extends down from the mountains, like a lawn. Following up the main valley we reach Castle Rock, with its immense hay ranches and fortress-butte, and beyond is Larkspur, named after one of the most striking of the plains birds. Thence the run is through a section of billowy plains or depressed foot-hills, up a steady ninety-feet grade to the Divide, in whose vicinity we encounter a succession of high buttes and mesas, the lower portions being composed of sandstone, while the tops are of igneous rock or lava. These constantly suggest artificial forms of towers, castles and fortifications, in some places rising nearly a thousand feet above the railway. Not infrequently the cliffs are so regularly disposed that it is hard to believe them merely natural formations. The entire scenery of this great ridge, and extending far out into the plains, is of an unique and interesting character. Near the summit there are remarkable evidences of its having been the coast-line of an ancient sea. The streams which rise on the northern slope of this watershed find their way into the Platte, while those on the southern declivity flow into the Arkansas. The Divide has a good covering of pines, often arranged by nature with park-like symmetry, and forming a charming contrast with the bare but beautifully colored cliffs. This region has been a chief source of Denver's lumber supply, and the timber tract is estimated to contain about 70,000 acres.

On the Divide is a beautiful sheet of water, known as Palmer Lake, from which is derived a large quantity of the purest ice. Here a novel and attractive depot has been erected by the railway company, and extensive improvements, including a dancing pavilion and summer hotel, cottages and boat-houses, have been made. In the hottest seasons the temperature is always cool and invigorating, and no spot within accessible distance is so well adapted for an economical and delightful resort for Denverites. On the southern face of the hill the rock-formations break out into still more marked resemblances to ruined castles, showing moats, arches and turrets. It follows that our Artist was enraptured with the romantic features of the place, and the Photographer insisted on taking out his camera and getting at work. One of his results, Phebe's Arch, is contributed to the pictorial fund of this volume. The descent from the Divide is rapid, and our attention is absorbed by the swiftly changing panorama. Close by are the mountains—their snowy pyramids holding entranced your eyes from far, far out on the plains. "In variety and harmony of form," said Bayard Taylor of them: "in effect against the dark blue sky, in breadth and grandeur, I know of no external picture of the Alps which can be placed beside it. If you could take away the valley of the Rhone, and unite the Alps of Savoy with the Bernese Oberland, you might obtain a tolerable idea of this view of the Rocky Mountains." Pike's Peak is constantly in sight, and every curve in the steel road presents it in a different aspect.

IN QUEEN'S CAÑON.

Presently we find ourselves on the bank of Monument Creek, pass the station of the same name, and soon encounter a series of small basins, and side valleys, green-carpeted and with gently sloping and wooded sides.

"Observe those odd rocks!" suddenly exclaims the Madame. "Notice how, all along the bluff, stand rows of little images, like the carved figure-friezes of the Parthenon; and how those great isolated rocks have been left like the discarded and broken furniture of Gog and Magog."

"Yes," I say, "but the tone of your imagery is low. Long, long ago a higher sentiment called them 'monuments,' and this whole illy-defined region of grotesquely-cut sandstones, Monument Park."

And then we all fall into a discussion of the process of formation of these quaint obelisks, which is interrupted by the Artist.

"Here is some pertinent testimony in Ludlow's admirable book, the 'Heart of the Continent,' which by your leave I will read to you. Ready?"

"Fire away!" we reply, and do the same with our cigars, making a treaty of amity in the blaze of a mutual match.

"'I ascended one of the most practicable hills among the number crowned by sculpturesque formations. The hill was a mere mass of sand and débris from decayed rocks, about a hundred feet high, conical, and bearing on its summit an irregular group of pillars. After a protracted examination, I found the formation to consist of a peculiar friable conglomerate, which has no precise parallel in any of our Eastern strata. Some of the pillars were nearly cylindrical, others were long cones; and a number were spindle-shaped, or like a buoy set on end. With hardly an exception, they were surmounted by capitals of remarkable projection beyond their base. These I found slightly different in composition from the shafts. The conglomerate of the latter was an irregular mixture of fragments from all the hypogene rocks of the range, including quartzose pebbles, pure crystals of silex, various crystalline sandstones, gneiss, solitary horn-blende and feldspar, nodular iron stones, rude agates and gun-flint; the whole loosely cemented in a matrix composed of clay, lime (most likely from the decomposition of gypsum), and red oxide of iron. The disk which formed the largely projecting capital seemed to represent the original diameter of the pillar, and apparently retained its proportions in virtue of a much closer texture and larger per cent. of iron in its composition. These were often so apparent that the pillars had a contour of the most rugged description, and a tinge of pale cream yellow, while the capitals were of a brick-dust color, with excess of red oxide, and nearly as uniform in their granulation as fine millstone-grit. The shape of these formations seemed, therefore, to turn on the comparative resistance to atmospheric influences possessed by their various parts. Many other indications . . . . led me to narrow down all the hypothetical agencies which might have produced them, to a single one, —air, in its chemical or mechanical operations, and usually in both. . . . One characteristic of the Rocky Mountains is its system of vast indentations, cutting through from the top to the bottom of the range. Some of these take the form of funnels, others are deep, tortuous galleries known as passes or cañons; but all have their openings toward the plains. The descending masses of air fall into these funnels or sinuous canals, as they slide down, concentrating themselves and acquiring a vertical motion. When they issue from the mouth of the gorge at the base of the range, they are gigantic augers, with a revolution faster than man's cunningest machinery, and a cutting-edge of silex, obtained from the first sand heap caught up by their fury. Thus armed with their own resistless motion and an incisive thread of the hardest mineral next to the diamond, they sweep on over the plains to excavate, pull down, or carve in new forms, whatever friable formation lies in their way.'"

By this time Colorado Springs was at hand, and as we had decided, like all other sensible people who come to Colorado, to sojourn awhile there and at Manitou, our cars were side tracked. And while Amos betook himself to the preparation of our evening meal, we admired the gorgeous sunset, and disposed our effects for the first night out.

Henry Ward Beecher once said that while the new birth was necessary to a true Christian life, it was very important that one be born well the first time. Colorado Springs was born well. It was organized on the colony plan, and the first stake was driven in July, 1871. Intelligent and far-seeing men were leaders of the enterprise, and in no way was their sagacity more apparent than in the insertion, in every deed of transfer, of a clause prohibiting, upon pain of forfeiture, the sale or manufacture of alcoholic beverages on the premises conveyed. This temperance clause was introduced, by General W. J. Palmer, the president of the colony, who during his services as engineer of railway extensions, had observed the destruction which the unrestrained traffic in intoxicants worked to life and property. It was not sentiment, but a sound business precaution, as the result has proved. Of course this provision has been contested, but it has been legally sustained, and has given the town the best moral tone of any in Colorado. The location was also wisely chosen, broad and regular streets were carefully laid out, a system of irrigation established, thousands of trees planted, and reservations for parks set aside. Some of the avenues running north and south might with propriety be designated boulevards, being 140 feet in width, with double roadways separated by parallel rows of trees. Other trees shade the walks at either side, and at their roots flow rapid streamlets of clearest water. The drives are smooth and hard, and the soil never becomes muddy, the moisture penetrating rapidly through the light gravelly loam. The gentle inclination southward renders drainage a very simple matter.

Seen from the railway, the town appears to be located upon a considerable elevation. In fact it stands upon a plateau in the midst of a valley. The thirty-five miles of streets and avenues are closely lined with substantial business blocks, pretentious residences, or tasty cottages. The pink and white stone of the Manitou quarries is largely used; and pent-roofs, ornamental gables, red chimneys, and the whole category of *renaissance* peculiarities, have representation in the architecture. The dwellers in these abodes are principally of the cultured and refined classes. Invalids from the intellectual centers of the East find health and congenial society here, while numbers of opulent mine owners and stockmen make the Springs their winter home.

The public buildings are all creditable; the Deaf-Mute Institute, Colorado College, the churches and schools being specially noteworthy.

CHEYENNE FALLS.

The Opera House is a veritable *bijou*, handsome and convenient in all its appointments, and, with a single exception not surpassed west of the Missouri. The new hotel, The Antlers, erected at a cost of over $125,000, is of stone, and is without doubt the most artistic and elegant structure of its kind in the State. It occupies a sightly position at the edge of the plateau, and from its balconies and verandas a marvelous and most inspiring view is presented. The foothills lie along the west, about five miles distant, the massive outlines of Cheyenne Mountain a little to the left, and the huge red towers that mark the gateway to the Garden of the Gods lifting their crests over the Mesa at the right, while above them all is reared the snow-crowned summit of Pike's Peak. To the north, is seen in the foreground the gray shoulders of the buttes, and in the distance the dark pine-covered elevation of the Divide. Easterly the land rises gently in a gray, grass-clad plain, until it cuts the blue horizon with a level line; while southward the mountains trend away, purple in the distance.

Colorado Springs lies under the shadow of Pike's Peak; and in the short autumn days the sun drops out of sight behind the mountain with startling suddenness at four o'clock. Then come the cool shadows, when fires have to be replenished, and doors and windows closed. From ten o'clock until the sun hides behind the hills, the blue skies, the soft breezes the grateful warmth, suggest that month in which, if ever, come perfect days. The June roses are absent, but the days are as rare as a day in June. The average temperature here is sixty degrees, and there are about three hundred days of sunshine in the year.

Within a radius of ten miles about the Springs are to be found more "interesting, varied and famous scenic attractions than in any similar compass the country over," we are told by the guide, and we are quite ready to believe when they are recounted. A drive of three miles across the Mesa, with its magnificent mountain view, brings you to Glen Eyrie, the secluded home of General Palmer, originator of the Denver and Rio Grande railway. "At the entrance you pass a little lodge — a sonnet in architecture, if one may so express it — the small but perfect rendering of a harmonious thought; you cross and recross a rushing, tumbling mountain brook over a dozen different bridges, some rustic, some of masonry, but each a gem in design and fitness; then at last, after the mind is properly tuned, as it were, to perfect accord, the full symphony bursts upon you. In the shadow of the eternal rock, with the wonderful background of mountains, surrounded by all that art can lend nature, is this delicious anachronism of a Queen Anne house, in sage-green and deep dull red, with arched balconies under pointed gables, and carved projections over mullioned windows, and trellised porches, with stained glass loopholes and an avalanche of roofs." A little distance from the house strange forms of red sandstone lift their heads far above the foliage, like a file of genii marching down on solemn mission from their abodes in mountain caves, while on the ledges of the gray bluffs opposite the eagles have built their nests. Farther up the Glen, and yet a part of it, is Queen's Cañon, a most rugged gorge, in which the wildness of nature has been for the most part unopposed. The same turbulent brook comes dashing down, in a series of cascades and rapids, from the Devil's Punch-Bowl, near the head of the Cañon. Rustic bridges cross it near the foot one of which is made the subject of an engraving; but soon the pathway breaks into a mere trail, which leads over boulders and fallen tree-trunks, or clings to precipitous cliffs which tower high overhead.

One mile north of Glen Eyrie is Blair Athol, with its exquisitely tinted pink sandstone pillars; while about the same distance to the south is the Garden of the Gods, which it seems, however, more proper to classify with Manitou's environs. Five miles northeasterly from the Springs are Austin's Bluffs, and a few miles west of these, Monument Park. Nearer by, and due west, are the Red and Bear Creek Cañons. An excellent way of reaching Pike's Peak is by the Cheyenne Mountain toll road, which terminates in a good trail passing the Seven Lakes. The Cheyenne Cañons, at the northern base of the mountain of the same name, are greatly frequented, and justly rank high in the category. They are two in number.

South Cheyenne Cañon is full of surprises. "The vulgar linear measure of its length is out of harmony with the winding path over rocks, between straight pines, and across the rushing waters of the brook that boils down the whole rocky cut. The stream, tossing over its rough bed and dropping into sandy pools, drives one from side to side of the narrow passage-way for foothold. Eleven times one crosses it, by stepping from one rolling and uncertain stone to another, by balancing across the lurching trunk of a felled tree, or by dams of driftwood; and, finally, skirting a huge boulder that juts out into the water, and jumping from rock to rock, the head of the Cañon is reached. The narrow gorge ends in a round well of granite, down one side of which leaps, slides, foams and rushes a series of waterfalls. Seven falls in line drop the water from the melted snow above into this cup. Looking from below, one sees (as in our illustration) only three, that, starting down the last almost perpendicular wall and striking ledges in the rock, and oblique crevices, send their jet shooting in a curved spray to the pool. In this deep hollow only the noonday sun ever shines, and a narrow bank of snow lies against one side in the shadow of the cliff. Going up the Cañon, with the roar of the waters ahead and the wild path before one, the loftiness and savage wildness of the walls catch only a dizzying glance, but coming out, their sides seem to touch the heavens and to be measureless. The eye can hardly take in the vast height, and with the afternoon sun touching only the extreme tops, one realizes in what a crevice and fissure of the rocks the Cañon winds. Across the widest place between the walls a girl could throw a stone, and from that it narrows even more. The cool, dim light down at the base contrasts strangely with the red blaze that reflects from the top of the high walls; and emerging from one group of pine trees, a turn in the Cañon confronts one with a whole wall of sandstone burning in the intense sunlight. A comparison between this and the Via Mala and the other wild gorges of the Alps is impossible, but had legend and history and poetry followed it for centuries, South Cheyenne Cañon would have its great features acknowledged. Let a ruined tower stand at its entrance, whence robber knights

IN NORTH CHEYENNE CAÑON.

had swooped down upon travelers and picked out the teeth of wealthy Hebrews; let a nation fight for its liberty through its chasm; and then let my Lord Byron turn loose the flood of his imagery upon it. After all, its wildness and untouched solitude are most impressive; and without history, save of the seasons, or sound, save of the wind, the water and the eagles, centuries have kept it for the small world that knows it now." So discourses a very charming lady writer.

North Cheyenne Cañon is scarcely less interesting, though less widely known. Its beauty is of a milder type, the walls advancing and retreating, and anon breaking into gaily-colored pinnacles on which the sunlight plays in strange freaks of light and shade. The little brook winds like a silver band beside the path, encircling the boulders which it cannot leap, and all the while singing softly to the rhythm of swaying vines. The birds chirp in unison as they skip from rock to rock, and in the harmony and essence of the scene all are subdued—save the Artist, whose deft pencil cannot weary in so much loveliness. When words fail, it is fortunate he is at hand to rescue writer and reader alike.

It was during our stay at Colorado Springs we made acquaintance with the burro. It is the nightingale of Colorado; its range of voice is limited, consisting indeed of only two notes; but the amount of eloquence, the superb quality, the deep resonance and flexible sinuosity which can be thrown by this natural musician into such a small compass are tremendous. As they lope down the street, the larboard ear in air, while the starboard droops limply, the long tapir-like nose quivering with the mighty volume of sound which is pouring through it, the sloping Chinese eyes looking at you sideways with the lack-lustre expression of their race, and an artistic kick thrown in occasionally to produce the tremolo that adds the last touch of grace to the ringing voice, you are overwhelmed.

We betook ourselves to the train one evening, after our by no means thorough exploration of the neighborhood, and began our preparations for a few days' absence at Manitou. It was only five miles away, and we had decided not to take our cars up. Retiring early, we fell asleep to dream of new pleasures, for which our appetites were already whetted.

III

A MOUNTAIN SPA.

> . . . And the ray
> Of a bright sun can make sufficient holiday,
> Developing the mountains, leaves, and flowers,
> And shining in the brawling brook, whereby,
> Clear as its current, glide the sauntering hours
> With a calm languor, which, though to the eye
> Idlesse it seem, hath its morality.
>
> —PETRARCH

AS well omit the Lions of St. Mark from a visit to Venice, as to pass by Manitou in a tour of Colorado. Manitou, the sacred health-fountains of Indian tradition, the shrine of disabled mountaineers, the "Saratoga" of the Rockies.

Leaving Colorado Springs, a branch of the railway swings gracefully around the low hills in which the Mesa terminates, and points for the gap in the mountains directly to the west. Nearly three miles from the junction we pass the driving park, and immediately after run up to Colorado City, the first capital of the Territory, and now a quiet little hamlet, whose chief industry is the production of much beer. Recalling the temperance proclivities of the Springs, it is unfortunately not strange that a drive to Colorado City in the long summer evenings should prove so attractive to the ardent youth of untamed blood. Then the road passes into the clusters of cottonwoods and willows which fringe the brook, crossing, recrossing, and dashing through patches of sunlight, whence the huge colored pinnacles in the Garden of the Gods are descried over the broken hills at the right. It is a striking ride, and curiously the location of the railway has not been allowed to mar its native beauty. Describing the contour of a projecting foothill, we obtain our first glimpse of Manitou, with its great hotels, its cut-stone cottages, and its picturesque station. Just across the way, and in the lowest depression of the narrow vale, is a charming villa, embowered in shrubbery, with quaint gables and porches, and phenomenal lawns and flower-beds. It is the home of Dr. W. A. Bell, for years vice-president of the Denver and Rio Grande railway.

The village itself is grouped in careless ease along the steep and bushy slopes of the valley, straggling about and abounding in miniature châlets, precisely as a mountain village ought to do. The Fontaine-qui-

Bouille, full of the sprightliness of its youth in Ute Pass, and its escapade at Rainbow Falls, comes dashing and splashing, and singing its happy song:

> "I chatter over stony ways,
>   In little sharps and trebles;
> I bubble into eddying bays,
>   I babble on the pebbles."

Down close to this frolicsome, icy-cold stream, are built the larger hotels, the Beebee and Manitou, surrounded by groves of cottonwood, aspen, wild cherry and box elder. They are cheery, clean, homelike,

A GLIMPSE OF MANITOU AND PIKE'S PEAK.

and handsomely furnished. The broad piazzas afford the finest views of Pike's Peak, Cameron's Cone, and their *confreres*. Here gather the "beauty and chivalry" of many climes, and in the long, soft evenings, devoid of dew or moisture, the cozy nooks offer the seclusion for — we had nearly said, flirtation — or cool refuge from the heated dancing-hall. Rustic bridges cross the brook, leading into a labyrinth of shade and on up to the crags behind. At the rear of the hotels, Lover's Lane, a most romantic ramble, starts out in a half-mile maze through arbors, and flowering shrubs, and over little precipices, for the springs. Beside the path, and in out-of-the-way spots among the bushes, are alluring seats, only large enough for two, where you may sit, while at your feet the selfsame brooklet murmurs:

> "I steal by lawns and grassy plots;
> I slide by hazel covers;
> I move the sweet forget-me-nots
> That grow for happy lovers."

Further up stream, a little way, are the homes of the citizens, and more hotels and boarding places — the Cliff House, Barker's, and a dozen others. Here, too, on the banks of the creek, boiling up in basins of their own secretion, and hidden under rustic kiosks of a later date, are the springs themselves. They are six in number, varying in temperature from 43° to 56° F., and are strongly charged with carbonic acid. "Coming up the valley," writes an authority, "the first is the Shoshone, bubbling up under a wooden canopy, close beside the main road of the village, and often called the Sulphur Spring, from the yellow deposit left around it. A few yards further on, and in a ledge of rock overhanging the right bank of the Fontaine, is the Navajo (shown in the foreground of our picture), containing carbonates of soda, lime and magnesia, and still more strongly charged with carbonic acid, having a refreshing taste similar to seltzer water. From this rocky basin, pipes conduct the water to the bath-house, which is situated on the stream a little below. Crossing by a pretty rustic bridge, we come to the Manitou, close to an ornamental summer-house; its taste and properties nearly resemble the Navajo. Recrossing the stream and walking a quarter of a mile up the Ute Pass road, following the right bank of the Fontaine, we find, close to its brink, the Ute Soda. This resembles the Manitou and Navajo, but is chemically less powerful, though much enjoyed for a refreshing draught. Retracing one's steps to within two hundred yards of the Manitou Spring, we cross a bridge leading over a stream which joins the Fontaine at almost a right angle from the southwest; following up the right bank of this mountain brook, which is called Ruxton's Creek, we enter the most beautiful of the tributary valleys of Manitou. Traversing the winding road among rocks and trees for nearly half a mile, we reach a pavilion close to the right bank of the creek, in which we find the Iron Ute, the water being highly effervescent, of the temperature of 44° 3' F., and very agreeable in spite of its marked chalybeate taste. Continuing up the left bank of the stream for a few hundred yards, we reach the last of the springs that have been analyzed — the Little Chief; this is less agreeable in taste, being less effervescent and more strongly impregnated with sulphate of soda than any of the other springs, and containing nearly as much iron as the Iron Ute.

"These springs have from time immemorial enjoyed a reputation as healing waters among the Indians, who, when driven from the glen by the inroads of civilization, left behind them wigwams to which they used to bring their sick; believing, as they did, that the Good Spirit breathed into the waters the breath of life, they bathed and drank of them, thinking thereby to find a cure for every ill; yet it has been found that they thought most highly of their virtues when their bones and joints were racked with pain, their skins covered with unsightly blotches, or their warriors weakened by wounds or mountain sickness. During the seasons that the use of these waters has been under observation, it has been noticed that rheumatism, certain skin diseases, and cases of debility have been much benefited, so far confirming the experience of the past. The Manitou and Navajo have also been highly praised for their relief of old kidney and liver troubles, and the Iron Ute for chronic alcoholism and uterine derangements. Many of the phthisical patients who come to this dry, bracing air in increasing numbers are also said to have drunk of the waters with evident advantage.

"Professor Loew (chemist to the Wheeler expedition), speaking of the Manitou Springs as a group, says, very justly, they resemble those of Ems, and excel those of Spa — two of the most celebrated in Europe.

"On looking at the analyses of the Manitou group it will be seen that they all contain carbonic acid and carbonate of soda, yet they vary in some of their other constituents. We will, therefore, divide them into three groups of carbonated soda waters: 1. The carbonated soda waters proper, comprising the Navajo, Manitou and Ute Soda, in which the soda and carbonic acid have the chief action. 2. The purging carbonated soda waters, comprising the Little Chief and Shoshone, where the action of the soda and carbonic acid is markedly modified by the sulphates of soda and potash. 3. The ferruginous carbonated soda waters where the action of the carbonic acid and soda is modified by the carbonate of iron, comprising the Iron Ute and the Little Chief, which latter belongs to this group as well as to the preceding one."

Such are the medicinal fountains that not only have proved themselves blessings to thousands of invalids, sick of pharmacy, but cause the summer days here to be haunted by pleasure-seekers, who make the health of some afflicted friend, or weariness from overwork in themselves, excuse for coming; or boldly assert themselves here purely for pleasure. Time was when this entrance to a score of glens was a rendezvous for game and primitively wild. Even a dozen years ago it would answer to this description, and now one need not go far in winter to find successful shooting. "In summer time," to quote the Earl of Dunraven, "beautiful but dangerous creatures roam the park. The tracks of tiny little shoes are more frequent than the less interesting, but harm-

THE MINERAL SPRINGS.

less, footprints of mountain sheep. You are more likely to catch a glimpse of the hem of a white petticoat in the distance than of the glancing form of a deer. The marks of carriage wheels are more plentiful than elk signs, and you are not now so liable to be scared by the human-like track of a gigantic bear as by the appalling impress of a number eleven boot."

Do not imagine, however, that all the boot-tracks mark "the appalling impress of a number eleven." The Madame tells me they dress as well at Manitou as at Saratoga; to me this seems a doubtful kind of compliment, but she intended it to cover the perfection of summer toilets. At Manitou, indeed, you do all you think it proper to do in the Green Mountains, or at White Sulphur, or any other upland resort, but in far more delightfully unconventional ways, and the enjoyment is proportionately increased.

No Eastern watering-place affords opportunity for so many desirable excursions, each distinct from the other in interest, each superb and of itself a sufficient inducement to come to Colorado. Just overhead towers the glorious old crest of Pike's Peak, the beacon of '59, and ever since the type of American mountains. He who does not ascend the Peak (if he is in fair health) can never get a good character from

Manitou. Of course all of the present party went. Moreover we went fancy-free and note-book forgotten—a happiness as great as Patti's when she saw there was no piano on the ocean steamer in which she was to take passage. "How was this?" do you ask? Lillian Seidmore had been there before us, and reaped with her keen sickle every spear of wheat in the whole field. To show our gleanings would amount to nothing; so here is her whole sheaf:

"The tenth of June having left the world upon its axis, a little band of heroic spirits made ready to mount the bony bronchos, and toil upward from the green and lovely vale of Manitou to the rocky height above. The noonday sun was sending down its most scorching rays, and the idlers on the hotel piazza were mopping their brows and repeating the wearisome formula of 'the hottest day ever known in Colorado.' The sun was ardent, to say the least, but the crisp breeze that came rustling down from the higher cañons tempered its effects.

"The sympathetic chambermaid of the Beebee House had been hovering in my doorway for a half hour before the start, urging me to take more and more wraps, and relating horrible anecdotes of the Chicago lady 'who had her nose burned to a white blister and her face so raw, ma'am, that we could hardly touch it with a feather for three days.' With such gentle admonition there was no struggle when the kind-hearted one proceeded to apply her preventive, and under a triple layer of cold cream, powder and barege veils we made the trip, and returned rather fairer in skin for the bleaching process. The perspiration ran off the guide's forehead before he had strapped on the first bundle of overcoats, ulsters, shawls, rugs and furs, but the grateful sensation they imparted to us a few hours later will cheer me through many midsummer days. The party included, among others, a gentleman and his wife from St. Louis, and the same wicked Colorado editor who is the author of all the fine spun yarns about the Pike's Peak volcano and the mountain lions.

"Such horses as we rode can be raised and trained only on a mountain trail, and if they could but speak, what tales of timidity, stupidity and absurdity they might relate. My own Arabian was a tan-colored beast, shading off to drab and old gold, known in the vernacular of the country as a buckskin horse and rejoiced in the sweet name of 'Bird.' It was a veritable misnomer, for birds do not generally sit down and roll at every piece of green grass or cool water that they come to, nor try to shake their riders off over their necks. My sudden flights to earth were heralded in all the turgid and flamboyant rhetoric of the circus ring, and equestrian feats, each outrivaling the other in novelty and unexpectedness, diversified the route. It was proposed to call the creature Jordan, because she rolled; and again it was suggested that as it was 'sinched' out of all shape it had mistaken itself for an hourglass, and concluded that it was time to turn. Another horse for a lady

2*

rider answered to 'Annie,' and this gentle beast was only kept from lying down in every stream by energetic pullings and vigorous thrashings. The good son of St. Louis, bidden, like Louis XVI at the guillotine, to 'mount to heaven,' when he leaped upon his dappled gray, in a linen coat, broad-brimmed hat and full-spread umbrella, had a truly ministerial air as he preceded the line up the road. The editor rode a pensive nag that hung its head and coughed timidly now and then, but chirruped to as 'Camille' would push forward and crowd the other horses off the trail, until a kicking and lashing from the heels of 'Bird' brought things in order.

"The Pike's Peak trail is one series of picturesque surprises. All that green cañons, tremendous boulders and turbulent little streams can do for beauty are there, and from the rustic spot where a small bandit on the rock demanded toll, there was a succession of grand and lovely scenes. The trail, worn deep into the grassy places by the procession of horses that goes up and down it from May to October, winds on between great rocks, along the steep and dizzy sides of cañons, past cascades and waterfalls (one of which is the subject of a sketch), and continually upward, opening boundless views out upon the broad plains that stretch like a yellow sea from the foothills of the Peak eastward. With every rise there came a greater one beyond, and above it all, seeming to move and rise further and higher from us, was the rose-red summit, with streaks and patches of snow bringing out its beautiful colors. Over giant boulders, creeping a cramped path beside and under them, or along a narrow ledge of sliding sand with colossal rocks miraculously suspended above our pathway, the panting horses toiled along. Ascending into higher and rarer air it was necessary every few minutes to stop and give the poor creatures a chance to breathe.

"As we rose higher on the mountain side more extended views were opened backward over the plains. The lowering sun fell fiercely on the red sandstone gateways of the Garden of the Gods, until they burned in flame-colored light against the yellow-gray grass. The hotels and cottages of Manitou were tiny dots in a green hollow far below, and the courses of the winding streams could be traced for miles over the plains by their green borders of cottonwood and willow trees. Wild flowers grew luxuriantly all the way, and in a little park half way to the summit, where the guides rest by a spring and wait for ascending and descending parties to pass, the ground was thick with big columbines, wild roses, harebells, white daisies, pale lavender geraniums with their petals streaked with maroon, and the beautiful blue-eyed penstemon of early June. At timber-line the wild box covered the sandy slopes with a thick and tangled mat of green, and higher than the hardiest pines stretched a rolling mountain meadow, a mile of emerald turf jeweled with the brilliant blossoms of bluebells, buttercups, dwarf sunflowers and dainty little Quaker-lady forget-me-nots.

"Sixteen people passed us in the half-way park on their way down. The terrified countenance of one lady on a mule would have made the hard-hearted to laugh. She pitched back and forth in her saddle, and shot a pitiful gaze at us as she went by that plainly indicated her estimate of us and mountain climbers in general. The twelve miles of steep, hard riding to the summit is trying to the most practiced rider; and for women, who have never sat a horse before, to attempt to make the trip up and down in one day is a folly that fully deserves the punishment it gets. Twenty-four miles of horseback riding on a level road even is apt to be remembered by the inexperienced. Added to the fatigue is the sea-sickness consequent upon the great altitude, and few who make the ascent escape that ill. It is a certificate of a rock-bound constitution to spend a night on the summit and not be grievously ill. After the mountain meadow come three miles of broken and ragged rock, the most wearisome and discouraging part of the road. The horses' sides throbbed frightfully, the keen winds made a halt for overcoats necessary, and the scramble over these steep rocks is a fearful thing in a nipping sunset breeze. The rocks of the summit, that seem only reddish brown from below, are of the softest pink and rose-red shades, dotted with black and golden moss-patches until they strongly remind one of the exquisite colors of speckled trout. Above this sea of loose and broken granite a low, square house of stone at last arose, and over the ultimate rock we finally stood on the highest inhabited point on the continent.

"The officer of the signal service, who lives in that lofty house, stood in his doorway shooting at a tin can on a pole, and in that thin open air the pop of the pistol was a short, faint little noise without crash or echo. The red ball of the sun sinking down behind the snowy edges of the mountains beyond Leadville sent strange lights and mists across the tossed and uneven stretch of mountains and parks that lay between it and the gaunt old Peak. The seventy acres of wildly scattered rock-fragments that crown the top afford a vantage ground for views to every point of the compass. Eastward across the vast prairie land there seems no limit to the vision, and beyond the green lines of the Platte and Arkansas rivers we amused ourselves by imagining the steeples of St. Louis in the rose and purple vapors of the horizon. The clouds, mists, shadows and faint opalescent lights on the plains, shifting, changing and fading each moment, are more fascinatingly beautiful than the dark, upheaved and splintered ridges of the mountains. Stretching out over the plains, at first in a blue cone upon the grass, and then sweeping outward and upward to the sky-line, the vast shadow of the mountain was thrown sharply against the sky.

"Wrapped in furs and bundled in all the woolen warmth of heaviest winter clothes, the chill air of evening penetrated like a knife-edge, and we sat shivering on the rocks with pitiable, pinched and purple faces and chattering teeth. The afterglow in the east, when the sky and the plains melted in one purple line and a band of rose-color went up higher and higher, was more lovely even than the pure crimson and gold and blue of the sunset clouds.

"Around the crackling fire in the station we thawed our benumbed fingers and watched the observations taken from the various instruments and sent clicking off on the telegraph wires to Washington headquarters. The sergeant wound the alarm clock to rouse us at four o'clock the next morning, and, giving up the one sleeping-room to the ladies, retired with the gentlemen of the party to a bed of buffalo-robes in the kitchen. The awful stillness, the stealthy puffs of wind, and the sense of isolation and remoteness, were distressing at first; but the tobacco-laden air dulled us to sleep. As the fire died out, dreams of Greenland—glaciers and giddy snow-banks on impossible summits—seized and held us, until a shivering voice gave the alarm: 'It is all red in the east.'

"We had climbed all those miles purposely to see the spectacle of dawn, but there was unhappiness among the pinched and pallid enthusiasts who crept out on the rocks and watched the half-light on the plains deepen. A pale and withered moon hung overhead, and miles away on the plain lay a vast white cloud like a lake, until the rising sun touched it and sent it rolling and tossing like angry waves. A crimson ball sprang suddenly from the outermost rim of the earth, glared with a red and sleepy eye upon the world, and pulled the cover of a cloud above it for a second nap before it came forth in full splendor. The shadow of the Peak projected westward fell this time on the uneven mountains, whose sides and clefts were filled and floating with faint pearl, lilac and roseate mists. The black patch where Denver lay on the plains, the snowy top of Gray's Peak, the green basin of South Park, and seemingly everything from end to end of the State, could be seen. Shivering, freezing, on that mountain top, with a fur cloak about me, besides all the other wraps, it seemed that there never was a winter day half as cold.

"In all the crevices of the rocks, wherever there was enough powdered granite to form a soil for their roots, were tiny little white blossoms, fairy stars or flowers, with just their heads above the ground, and an exquisite perfume breathing from them. Bidding the guide to sinch up quickly for the down trip, we partook of the signal sergeant's coffee, and listened to his anecdotes of his lonesome life of two weeks on the mountain and two weeks in town.

"'You are the best crowd that's been up,' said the brave man of barometers. 'They all get sick when they stay over night. It took ME a month to get used to it. You ought to stay until noon and see the tender-feet come up and get sick. Oh, Lord! there was an old lady up here the other day, and she says to me: "Sergeant, don't people ever die of this sickness up here?" "Oh, yes, ma'am," says I, "a lady died

PIKE'S PEAK TRAIL.

the other day, and as there wasn't any one to identify her we just put her over in that snow-bank there."'

"With a lot more of such mountain horrors he kept his rafters ringing, and then bade us climb the ladder to the top of his house, which would make up the difference of fifteen feet between his abode and Gray's Peak. We looked at the grave of the imaginary child destroyed by mountain rats, gave a last glance at the enchanted view, and left the chilling region."

Another entertaining jaunt is a couple of miles or less up the Ute Pass wagon-road to Rainbow Falls, one of the finest cascades in the

West—where such things are more of a curiosity than in wetter regions of the world. The water comes down here with a more than ordinarily desperate plunge, and it is great sport to climb about the angular rocks that hem it in.

Ute Pass leads over into South Park, and before the days of railways it was greatly traveled by passengers and by freight wagons to Leadville and Fairplay. There is less transit there now, but in summer pleasure-parties constantly traverse the Pass, partly for its own sake and partly to enjoy a sight of Manitou Park on the opposite side, whence a magnificent array of the snowy interior ranges is to be seen, northward and westward, while Pike's Peak presents itself to superior advantage from that point of view. In the park is a good little hotel and dairy, and a trout stream and pond where the Eastern brook-trout has been assiduously cultivated. In the fall Manitou Park is the resort of deer hunters and grouse shooters.

Then there is the already mentioned Garden of the Gods, hidden behind those garish walls of red and yellow sandstone, so stark and out of place in the soberly-toned landscape that they travesty nature, converting the whole picture into a theatrical scene, and a highly spectacular one at that. Passing behind these sensational walls, one is not surprised to find a sort of gigantic peep-show in pantomime. The solid rocks have gone masquerading in every sort of absurd costume and character. The colors of the make-up, too, are varied from black through all the browns and drabs to pure white, and then again through yellows and buffs and pinks up to staring red. Who can portray adequately these odd forms of chiseled stone? I have read a dozen descriptions, and so have you, no doubt. But one I have just seen, in a letter by a Boston lady, is so pertinent you should have the pleasure of reading it:

"The impression is of something mighty, unreal and supernatural. Of the gods surely—but the gods of the Norse Walhalla in some of their strange outbursts of wild rage or uncouth playfulness. The beauty-loving divinities of Greece and Rome could have nothing in common with such sublime awkwardness. Jove's ambrosial curls must shake in another Olympia than this. Weird and grotesque, but solemn and awful at the same time, as if one stood on the confines of another world, and soon the veil would be rent which divided them. Words are worse than useless to attempt such a picture. Perhaps if one could live in the shadow of its savage grandeur for months until his soul were permeated, language would begin to find itself flowing in proper channels, but in the first stupor of astonishment one must only hold his breath. The Garden itself, the holy of holies, as most fancy, is not so overpowering to me as the vast outlying wildness.

"To pass in between massive portals of rock of brilliant terra-cotta red, and enter on a plain miles in extent, covered in all directions with magnificent isolated masses of the same striking color, each lifting itself against the wonderful blue of a Colorado sky with a sharpness of outline that would shame the fine cutting of an etching; to find the ground under your feet, over the whole immense surface, carpeted with the same rich tint, underlying arabesques of green and gray, where grass and mosses have crept; to come upon masses of pale velvety gypsum, set now and again as if to make more effective by contrast the deep red which strikes the dominant chord of the picture; and always, as you look through or above, to catch the stormy billows of the giant mountain range tossed against the sky, with the regal snow-crowned massiveness of Pike's Peak rising over all, is something, once seen, never to be forgotten. Strange, grotesque shapes, mammoth caricatures of animals, clamber, crouch, or spring from vantage points hundreds of feet in air. Here a battlemented wall is pierced by a round window; there a cluster of slender spires lift themselves; beyond, a leaning tower slants through the blue air, or a cube as large as a dwelling-house is balanced on a pivot-like point at the base, as if a child's strength could upset it. Imagine all this scintillant with color, set under a dazzling sapphire dome, with the silver stems and delicate frondage of young cottonwoods in one space, or a strong young hemlock lifting green symmetrical arms from some high rocky cleft in another. This can be told, but the massiveness of sky-piled masonry, the almost infernal mixture of grandeur and grotesqueness, are beyond expression. After the first few moments of wild exclamation one sinks into an awed silence."

The reader must see for himself these grotesque monuments, these relics of ruined strata, these sportive, wind-cut ghosts of the old *régime* here, these fanciful images of things seen and unseen, which stand thickly over hundreds of acres like the moldering ruins of some half-buried city of the desert, if he would fully understand.

Out of the many other sources of enjoyment near Manitou, the visitor will by no means neglect the Cave of the Winds. Though you may ride, if you wish, it is just a pleasant walk up Williams' Cañon, one of the prettiest of the gorges that seam the rugged base of the great Peak. The walls are limestone, stained bright red and Indian yellow, lofty, vertical and broken into a multitude of bastions, turrets pinnacles and sweeping, hugely carved façades, whose rugged battlements tower hundreds of feet overhead against a sky of violet. At their bases these upright walls are so close together that much of the way there is not room for one carriage to pass another, and the track lies nearly always in the very bed of the sparkling brook. You seem always in a *cul de sac* among the zigzags of this irregular chasm, and sometimes the abundant foliage, rooted in the crevices above, meets in an arch across the brightly-painted but narrow space you are tortuously threading.

Half a mile up the cañon, at the end of the roadway, a trail goes by frequent turnings up the precipitous sides of the ravine to where a sheer cliff begins, about three hundred feet higher. Floundering up this steep and slippery goat-path, we arrived breathless at a stairway leading through an arch of native rock into a great chimney, opening out to the sunlight above, and found opposite us a niche which served as ante-room and entrance to the cave.

The history of this cave is entertaining, for it was the discovery, in June, 1880, of two boys of Colorado Springs, who were members of an "exploring society," organized by the pastor of the Congregational Church there to provide the boys of his Sunday-school with some safe and healthful outlet for their adventurous spirits.

The cave, as we saw it, is a labyrinth of narrow passages, occasionally opening out into chambers of irregular size (and never with very high ceilings), into which protrude great ledges and points of rock from the stratified walls, still further limiting the space. These passages are often very narrow, and in many cases you must stoop in crowding through, or, if you insist upon going to the end, squirm along, Brahmin-like, on your stomach. The avenues and apartments are not all upon the same level, but run over and under each other, and constantly show slender fox holes branching off, which the guide tells you lead to some stygian retreat you have visited or are about to see. In remote portions of the cave there are very large rooms, like Alabaster Hall, some of which are encumbered with fallen masses and with pillars of drip stone.

The cave is not remarkable for large stalactites and stalagmites, but excels in its profusion of small ornaments, produced by the solution of the rock and its re-deposition in odd and pretty forms. From many of the ledges hang rows of small stalactites like icicles from wintry caves, and often these have fine musical tones, so that by selecting a suitable number, varied in their pitch, simple tunes can easily and very melodiously be played by tapping. In some parts of the cave, the stalactites are soldered together into a ribbed mass, like a cascade falling over the ledges. Elsewhere the "ribbon" or "drapery" form of flattened stalactites recalls to you the Luray Caves, though here it is carried out on a smaller scale; while in this particular, as in many others, reminding one of the magnificent Virginia caverns only by small suggestions, in one respect this cave far surpasses in beauty its Eastern prototype. The floors of many rooms are laid, several inches deep, with incrustations of lime work, which is embroidered in raised ridges of exquisite carving. Again, where water has been caught in depressions, these basins have been lined with a continuous, crowding plush of minute lime crystals, — like small tufted cushions of yellow and white moss. Such depressed patches occur frequently; moreover, the rapid evaporation of these pools, in confined spaces, has so surcharged the air with carbonated moisture, that particles of lime have been deposited on the walls of the pocket in a thousand dainty and delicate forms,— tiny stalactites and bunches of stone twigs,— until you fancy the most airy of milleporic corals transferred to these recesses. Here often the air seems foggy as

THE LEGEND OF THE CAVERN.

your lamp-rays strike it, and the growing filigree-work gleams alabaster-white under the spray that is producing its weird and exquisite growth. In this form of minute and frost-like ornamentation, the cave excels anything I know of anywhere, and is strangely beautiful.

This cavern, however, is sadly deficient in a proper amount of legendary interest. No human bones have been found, and no lover's leap has been designated. This misfortune must be remedied; and I have selected a dangerous kind of a place at which, hereafter, the following touching tradition will cause the tourist to drop a tear: Many, many years ago an Indian maiden discovered this cave while eagerly pursuing a woodchuck to its long home; the home proving longer than she thought, she crept quite through into the unsuspected enlargement of a cave-chamber, and a startled congregation of pensive bats. She told no one of her discovery, because she had not, after all, caught the woodchuck, and went without meat for supper. A noble warrior, who had done marvelous deeds of valor, loved the maiden. He wooed and she would, but the swarthy papa would n't. Sadness, anger, surreptitious trysting where the fleecy cottonwood waves melodiously above the crystal streamlet, etc., etc. The irate old warrior brings an aged brave, who has spent his whole life in doing nothing of more account than cronifying with the heart-sick girl's father. This man she must marry, and the young suitor must go. Refusals by the maiden, loud talk by the youth, sneers from the old cronies, flight of the lovers to the woodchuck's hole, vermicular but affectionate concealment, like another Æneas and Dido. The woodchuck, stealing forth, sees a wolf outside, trying to make him pay his poll-tax; so he sits quietly just inside his safe doorway, obscuring the light. Endeavoring to find their way about in the consequent darkness, the imprisoned lovers pitch headlong over the precipice I have referred to. Guide-books please copy.

Our train bore a pensive party down the valley of the Fontaine, as

RAINBOW FALLS.

it headed for Pueblo. The Musician drew a plaintive air from his violin, and as the friendly mountain range receded and dipped away in the West, we fell to wondering when, if ever, we should tread those vales again.

IV

PUEBLO AND ITS FURNACES.

In Steyermark—old Steyermark,
The mountain summits are white and stark;
The rough winds furrow their trackless snow,
But the mirrors of crystal are smooth below;
The stormy Danube clasps the wave
That downward sweeps with the Drave and Save,
And the Euxine is whitened with many a bark,
Freighted with ores of Steyermark.

In Steyermark—rough Steyermark,
The anvils ring from dawn till dark;
The molten streams of the furnace glare,
Blurring with crimson the midnight air;
The lusty voices of forgemen chord.
Chanting the ballad of Siegfried's Sword,
While the hammers swung by their arms so stark
Strike to the music of Steyermark!

—BAYARD TAYLOR.

IT is a fortunate introduction the traveler, fresh from the Eastern States and weary with his long plains journey, gets at Pueblo to the lively, progressive, *booming* spirit of Colorado. Here are the oldest and the newest in the Centennial State,—the fragments of tradition that go back to the thrilling, adventurous days of fur-trapping and Indian wars: the concentrated essence of later improvements; and the most practical present, mingled in a single tableau, for a telephone line crosses the ruins of the old adobe fort or Spanish "pueblo," which gave to the locality its name when it was an outpost for the traders from New Mexico

In its modern shape the town is one of the longest settled in the State, and a great flurry began and ended there years ago. Then, neglected by men of money, Pueblo languished and was spoken chidingly of by its sister cities in embryo. Now all this has changed, and, perhaps aroused by the prosperity of Leadville, Pueblo began about three years ago to assert herself, and to-day stands next to Denver in rank both as a populous and as a money-making center. No business man or statistician could find a more deeply entertaining study than the investigation of how this rejuvenation has arisen and been made to produce so striking results. Such an inquirer would find several large industries claiming to have furnished the turning point; but it is evident that the few who faithfully stood by the comatose town, and steadily struggled toward its commercial revival, were prompt to seize upon the altered flood and take advantage of the tide which led to fortune. The impression once advertised that Pueblo was shaking off her lethargy and about to become a second Pittsburgh, a thousand men of business were quick to catch the idea and make the "boom" a fact. Thus from 5,000 inhabitants in 1875 she has come to over 15,000 in 1883.

It is undoubtedly true that the Denver and Rio Grande railway has done more to aid this advancement than any other one agency; but an important impetus was given to Pueblo in 1878, when a company of gentlemen decided to build a smelter here. The work was put under the charge of its present superintendent, and ninety days from breaking ground the furnace was in operation. There was only a single small one at first; but fourteen are running now. Then there was a diminutive shed to cover the whole affair; now there are acres of fine buildings. Then a dozen men did all the work; now from 380 to 400 are employed, and the pay-roll reaches $375,000 per annum. That's the way they do things in Pueblo.

This smelter is on the northern bank of the river, just under the bluff. From a distance, all that you can discern over the trees is a collection of lofty brick and iron chimney-stacks, and wide black roofs. Coming nearer, the enormous slag-dump discloses the nature of the industry, and testifies to the quantity of ore that has passed through the furnaces. Though on the banks of a swift river, the works are run by steam, which can be depended upon for steady service, and on which winter makes no impression. A thousand tons of coal and seven hundred tons of coke a month are used, the cheapness and proximity of this fuel forming one of the inducements to place the smelter here. Cañon City and El Moro coals are mixed, but the coke all comes from the latter point. At the start an engine of 60-horse power supplied all needs, but a new one of 175-horse power has been found necessary, and Denver was able to manufacture it. As for the machinery, it is not essentially different from that in other smelters, except in small details, where the most approved modern methods are made use of. There are great rooms full of roasting-ovens, immense bins where the pulverized and roasted ore cools off, elevators that hoist it to the smelting furnaces, and all the usual appliances, in great perfection, for charging the furnaces, drawing off and throwing aside the slag, and for casting the precious pigs of bullion. All walls and floors are stone and brick; everywhere order and neatness prevail. This plant has already cost the firm $200,000, and they have enough more money constantly put into ore and bullion to make $750,000 invested at the works. The ore is bought outright, according to a scale of prices which is about as follows: Gold, $18.00 per ounce; silver, $1.00 per ounce; and copper, $1.50 per unit.

This is reckoned by "dry" assay, being two per cent. off from "wet." For the lead in the ore, 30 cents per unit up to 30 per cent., 40 cents up to 40 per cent., and 45 when over 40 per cent.; but both lead and copper will not be paid for in the same ore. From the total is deducted $20.00 as fee for treating it. For the assaying a capital laboratory of several rooms is provided, where two assayers and two chemists are continually busy. Every lot, as purchased, is kept separate and subject-

GARDEN OF THE GODS.

ed to a homeopathic process of dilution, until a sample is obtained that represents most exactly the whole. The arrangements for crushing and sampling the ore are very complete, and a large number of lots can be handled at once. When the final sample has been reached it is subjected to a very careful assay, not only to determine what shall be paid for it, but to find out what are its qualities in relation to the process of smelting. This process requires a certain percentage of lead, a certain amount of silica, and a certain proportion of iron and lime in each charge. It is the duty of the assayers and chemists to ascertain precisely the proportions of these ingredients in the ore under consideration, in order to know how much lead, iron, lime or silica, to add in order to make a compound suitable to fuse thoroughly, even to the dissolution of the desperately refractory zinc and antimony; and which, also, shall yield up every particle of silver and gold. The iron and lime have usually to be added outright in this smelter; but the proper proportions of lead and silica are obtained by combining an ore deficient in one of these elements, but containing an excess of the other, with ores oppositely constituted. It is one of the advantages of the smelter at Pueblo, that, being centrally placed and down hill from every mining district, it can draw to its bins ores of every variety; thus it is able to mix to the greatest advantage, and in this economy and the attendant thoroughness of treatment, lie the possibilities (and actuality) of profit.

The lime for flux is procured three miles from town, and costs only $1.00 a ton, while every other smelter in the State must pay from $2.50 to $5.00. Its building-stone is a splendid quality of cream-tinted sandstone procured in the mesa only a short distance away. So widely satisfactory have been its results that the Pueblo smelter does the largest business of any in Colorado. I saw ore there from away beyond Silverton; from Ouray and the Lake City region; from the Gunnison country and the Collegiate range this side; from Leadville (competitive with Leadville's own smelters); from Silver Cliff and Rosita; and, finally, from numerous camps in the northern part of the State. The last was surprising; for it meant that all that weight of ore had been brought hither past the furnaces at Golden and Denver, because the owners realized more for it, in spite of excess of freight, than they could get at home. The superintendent told me that they handled more ore from Clear Creek county than all the other smelters of the State; and he explained it by showing how his nearness to fuel, and consequent saving in this important item, with his cheap labor, permitted him to bid and pay more than any other smelter could for choice ores, for which a premium is given. These facts were held in mind when, encouraged by the railway, the smelter was placed here, and the expectations of its projectors have been more than gratified. The present capacity of the establishment is 125 tons of ore a day in the fourteen blast furnaces, and 100 tons a day in nine large calcining furnaces. Expensive improvements, and of the most solid character, are being made constantly in all parts of the works. The most important of these has been the erection of machinery for refining the bullion, which has a capacity of twenty-five hundred tons a month, and is constructed on the best known principles. It has been customary in the West to send to New York the lead which results from silver refining; it is made into sheet-lead and leaden pipe, in which form it is bought by wholesale houses in Chicago and St. Louis, to be again sent to Colorado, at the rate of

THE STEEL WORKS.

perhaps a thousand tons a year. Now it is proposed to keep at home the profits and freightage of this costly and heavy material in house construction. Machinery has therefore been added to make up all the lead into sheets, bars, and piping. This is done so cheaply, that Pueblo can now send it across the plains and undersell Chicago and St. Louis in the Mississippi valley. The supply largely exceeds the home demand, and a new export for this State has thus been created. Utah will henceforth yield a large portion of the bullion to be refined. Another experiment will be the refining of copper. There are various mines in New Mexico and Arizona—many of them worked in ancient days by the Spaniards—which supply a base form of metallic copper. This crude copper is now nearly all sent to Baltimore, and there refined and rolled. The New Mexico division of the Denver and Rio Grande will soon penetrate the region of some of these mines, while the Atchison, Topeka and Santa Fe makes others accessible. Side-tracks from both these roads run into the smelter's enclosure. To bring the copper here will therefore be an easy matter; and it can be produced in shape for commercial use much more cheaply than any Eastern factory is able to turn it out.

Another large factor in Pueblo's revival was the establishment there of the steel works. These are the property of the Colorado Coal and Iron company, composed of the leading men in the Denver and Rio Grande railway, so that, though the two corporations are distinct, their interests are closely allied. This powerful association was formed in 1879 by the consolidation of two or three other companies having similar aims, and it became the owner not only of the steel and iron works here, and of a great deal of real estate, but also of nearly all the mines of coal and iron now being developed in this State. Its capital is ten millions, and its principal offices are located at South Pueblo. It employs over two thousand men in its various enterprises, and is constantly enlarging its operations and perfecting its methods. Many of its coal banks will be referred to in other paragraphs of this volume. The ore derived from all its iron mines is exceptionally free from phosphorus, and therefore well adapted to the manufacture of steel. The subjoined analyses exhibit the character of these ores:

	Metallic Iron.	Silica.	Phosphorus.	Lime.	Alumina.	Magnesia.	Manganese.	Sulphur.
Placer	52.2	12.64	.051	5.70	3.6	3.12	.34	Trace.
Salida	65.8	5.78	.015	.34	1.5	.81	.22	.014
Villa Grove	57.3	5.03	.019	..	..	..	1.87	.006

A conservative estimate places the amount of iron ore the company has developed at over two millions of tons. Besides these high-grade ores, there are others of an inferior grade, which, being mixed with mill-cinders, will produce the commoner sorts of pig-iron, suitable for foundry-work. Limestone, valuable as flux, is quarried from a ledge within seven miles of the furnace, with which it is connected by rail, and the supply is practically inexhaustible. Gannister and fire-clay, also, are found in abundance in the vicinity. With coal, coke, iron and all the furnace ingredients radiated about this point, which, at the same time is nearest to the Eastern forges whence must be brought the massive machinery to equip the works, it requires no second thought to perceive that South Pueblo offers altogether the most profitable site for vast factories like these.

Immediately following this decision, in the spring of 1879, a large tract of mesa-land was secured, beside the track of the Denver and Rio Grande railway, about a mile south of the Union Depot, where not only the foundations of the mills, but a village-site was laid out and numerous side-tracks were put down. Very soon the tall chimneys of the blast furnaces began to rise into the ken of the people of Pueblo. Simultaneously a large number of fine cottages were built as homes for workmen, and other structures were set on foot, among them a commodious hospital, for joint use of the mills and the railway company. It is a very pleasant, well-ordered and growing little town, known as Bessemer, and even now the space between it and the city is rapidly filling up. The present daily output of the blast furnaces is one hundred tons of pig-iron, but soon a twin furnace will double the productive capacity of the works. Besides the furnaces, the plant includes Bessemer steel converting works, a rolling and rail mill, 450 by 60 feet, a nail mill, a puddling mill and foundry. All of these establishments are in every way equal to the best of their kind in the East. The blast furnace is fifteen feet base and sixty-five feet high, with fire-brick, hot-blast stoves, and a Morris blowing engine. In the steel-converting works the arrangement of the plant is similar to that of the new Pittsburgh Bessemer Steel Company, which has given exceptionally good results. The rail mill plant consists of Siemen's heating furnaces and heavy blooming and rail trains, and the puddling and nail mills are equipped with the best modern machinery. At Denver the same company owns a rolling mill, where bar and railroad iron and mine rails are manufactured; these in the future will mainly be supplied from South Pueblo. The effect upon Colorado of this forging of native iron for home consumption must be very important. All iron and iron ware is nearly doubled in value by the necessarily high rates of freight across the plains. Manufacturing, now that the crude material can be obtained on the spot, is cheaper than importing, and in the wake of the blast furnace must follow a long train of iron industries. Already negotiations are on foot for the establishment of extensive stove works (a million of dollars was sent East last year from Colorado in payment for stoves alone), and the erection of car-wheel shops is also contemplated. Indeed, in this mining country, which also is a region that is rapidly filling up with large towns, the demand for manufactured

STATISTICS OF PRODUCTION. 57

iron of all sorts is very large. It is another step in the gradual movement of trade-centers westward.

A still clearer idea of the great value of its interests to the State, and of its local works to Pueblo, may be obtained from the report of the productions of the Colorado Coal and Iron company for the year 1882 which, briefly summarized, aggregates as follows: Coal, 511,239 tons; coke, 92,770 tons; iron ores, 53,425 tons; merchant iron mine rails, etc., 3,883 tons; castings, 2,752 tons; pig-iron, 24,303 tons; muck bar, in four months, 1,253 tons; steel ingots, eight months, 20,919 tons; steel blooms, eight months, 18,068 tons; steel rails, eight months, 16,139 tons; nails, four months, 16,158 kegs; and spikes, six months, 5,022 kegs.

ENTRANCE TO CAVE OF THE WINDS.

The economy of location, and the successful results attending the establishment here of the great enterprises referred to, are attracting many others. During the past season one of Leadville's largest smelters, having been destroyed by fire, has been rebuilt at Pueblo, and more will naturally follow.

The mercantile part of the community, however, while admitting all the claims of the steel works and the smelter to their great and beneficent influence upon the destiny of the new town, puts forward its own claim to the credit of commencing the progressive movement. When, by the extensions of the railway into the back country of Colorado, merchants began to perceive that at Pueblo they could buy goods of precisely such grades as they desired a trifle more cheaply, and get them home a trifle more expeditiously, than by going to Denver or Kansas City, Pueblo began to feel the impulse of new commercial vigor. When it came to reckoning upon a whole year's purchases, the slight advantage gained in freight over Denver, to all southern and middle interior points, amounted to a very considerable sum. Here, far more than in the Eastern States, the freight charges must be taken into account by the country merchant; particularly in the provision business, where the staples are the heaviest articles, as a rule, and, at the same time, those on which the least profit accrues. This consideration, impressing itself more and more upon the good judgment of the mountain dealers, is bringing a larger and larger trade to Pueblo, until now she is beginning to boast herself mistress of all southern and middle Colorado and of northern New Mexico. She can not hope to compete with Denver for the northern half of the State, but she does not intend to lose her grip upon the great, rapidly-developing, money-producing San Juan, Gunnison and Rio Grande regions. And as the visitor sees the railway yards crowded with loaded freight trains destined for every point of the compass; notes the throng of laden carts in the furrowed streets; observes how every warehouse is plethoric with constantly changing merchandise, often stacked on the curbstone under the cover of a canvas sheet because room within doors can not be found; witnesses the temporary nature of so many scores of buildings for business and for domiciles, and learns how most of their owners are putting up permanent houses, multitudes of which are rising substantially on every side;—when he has caught the meaning of all this, he finds that Pueblo has an idea that her opportunities are great, and that she does not propose to neglect them in the least particular. There is much wealth there now, and more is being introduced by Eastern investors, or accumulated on the spot, not only in trade, but in the very extensive herds of cattle and sheep that center there. Yet this seems to be but the incipience of her prosperity,—a prosperity which rests on solid foundations, existing not alone in the industries I have catalogued and the trade which has centered there, but in the fact that values are not inflated and that the real property of the city is mortgaged to a remarkably small degree.

Pueblo, though I have treated it as a unit, really consists of two cities, having the rushing flood of the Arkansas between them. Each has water-works, and civic institutions separately, but I have no doubt this cumbersome duality will be done away with in time. It is on the South Pueblo side that the railways center at the Union Depot. The Atchison, Topeka and Santa Fe sends hither daily its trains from Kansas City, and hence the Denver and Rio Grande forwards its passengers northward to Denver, westward to Silver Cliff, Leadville, Salt Lake City and Ogden, and southward to El Moro, Alamosa, Del Norte, Durango, Silverton and Santa Fe. It is on this side, also, that the factories are, and that others will stand. Here, too, are being placed the great wholesale depots.

There is too much rush and dust and building and general chaos in the lowlands, where the business part of the town is, to make a residence there as pleasant as it will be a few years later; but upon the high mesa, whose rounded bluffs of gravel form the first break upon the shore of the great plains that extend thence in uninterrupted level to the Missouri River, a young city has grown up, which is admirable as a place for a home. Here are long, straight, well-shaded streets of elegant houses; here are churches and school buildings and all the pleasant appurtenances of a fine town, overlooking the city, the wooded valley of the Arkansas, the busy railway junctions, and the measureless plateaus beyond. One gets a new idea of the possibilities of a delightful home in Pueblo after he has walked upon these surprising highlands.

Nowhere, either, will you get a more inspiring mountain landscape— the far scintillations of the Sangre de Cristo; the twin breasts of Wahatoya; the glittering, notched line and clustering foothills of the Sierra Mojada; the great gates that admit to the upper Arkansas; and Pike's "shining mountain," surrounded by its ermined courtiers, only a little less in majesty than their prince.

V

OVER THE SANGRE DE CRISTO.

RALEIGH: Fain would I climb,
But that I fear to fall,
ELIZABETH: If thy heart fail thee,
Why, then, climb at all?

TOWARD the middle of one bright afternoon we were pulled out of Pueblo, our three cars having been attached to the regular south-bound express. We had fully discussed the matter, and determined to go on to the end of the track, or, more literally, to *one* end—for there are many termini to this wide-branching system—on to the warm old plazitas and dreamily pleasant pueblos of New Mexico. Why not?

But so inconsequential and careless an "outfit" was this, that no sooner had our minds been fairly settled to the plan (while the shadow of the Greenhorn came creeping out toward Cuchara and we were heading straight into its gloom) than somebody proposed our spending the night quietly in Veta Pass.

This mountain pass and its "Muleshoe," dwarfing in interest the celebrated "Horseshoe curve" in Pennsylvania, because occupying far less space, just as the foot of a mule may be set within a horse's track—these have been famous ever since railroading in southern Colorado began, and naturally we did not like to go past them in the darkness. We wanted to see how the track was laid away around the head of the long ravine, whether it doubled upon itself in as close a loop as they said it did; and whether the train really climbed through the clouds about the brow of Dump Mountain, as the pictures represented. So we told the conductor to drop us.

It was dark by the time we had been left in good shape on the terrace-like siding in Veta Pass, and, weary with our swift run, we were quite ready to shut out the gathering shades and be merry over our dinner; but first, all eyes must watch the departing express begin its climb up Dump Mountain. Think of swinging a train round a curve of only thirty degrees, on a very stiff grade, and with a bridge directly in the center of the turn! That is what this audacious railway does every few hours in the "toe" of the Muleshoe. From our lonely night-gripped dot of a house on the wild hill side, we could see squarely facing us both the Cyclopean blaze of the fierce headlight, and the two watchful red eyes glaring scornfully from the rear platform; by that we knew that the train had doubled on its own length of only six cars. Then, with hoarse panting and grinding of tortured wheels and rails, the two powerful locomotives began to force their way up the hill side right opposite us, the slanting line of bright windows showing how amazingly steep a grade of two hundred and seventeen feet to the mile really is. The beam of the headlight thrust itself forward, not level, as is its wont, but aimed at a planet that glimmered just above a distant ghostly peak.

"Do you remember," murmurs the Madame in a low voice, as she stands with bated breath beside me; "do you remember how Thoreau advises one of his friends to 'hitch his chariot to a star'? Doesn't this scene come near his splendid ambition? Will that train stop short of the sky, do you think?"

Surely it seemed that it would not, for only when the stokers opened the doors of the laboring furnaces, and volumes of red light suddenly illumined the overhanging masses of smoke, touching into strange prominence for an instant the rocks and trees beside the engines, could we see that the train stood upon anything solid, or was moving otherwise than as a slow meteor passes athwart the midnight sky. It was easy to imagine that long line of uncanny lights a fiery motto emblazoned upon the side of the dark mountain, and to read in it, *Per aspera ad astra.*

Yet the scene was far from fanciful. It was very real, and a fine sight for a man interested in mechanical progress, to watch those great machines walk up that hill, spouting two geysers of smoke and sparks, and dragging the ponderous train slowly but steadily along its upward course. Now and again they would be lost behind the fringe of woods through which the track passed, and then we would see the cone-like, rugged spruces sharply outlined against the luminous volumes of smoke. A moment later and the train disappeared around Dump Mountain, with a sardonic wink of the red guard-lights at the last; but presently we had knowledge of it again, for a fountain of sparks and black smoke from the engines blotted out the scintillating sky just above the highest crest of the ridge.

"Suppose it had broken in two on that hill-side," remarks the Artist, as I am carving the roast, five minutes later.

"It wouldn't have mattered," is the reply. "I once saw a heavier train than that break on this very mountain, the three rear coaches parting company with the forward portion of the train."

"But wasn't that criminal carelessness?" cries the Madame, who is death on inattention to railway duties. I should hate to be a neglectful brakeman before her gray eyes as judge!

"Not at all," she is informed. "The cars were properly coupled with what, for all any one could see, was a sufficiently strong link, but the strain proved too great for the tenacity of the iron."

"Well, I suppose they went down the track a flying, or else over the precipice."

"Not a bit of it. The two parts of the train stopped not more than twenty feet apart. When the accident occurred the engineer knew it instantly by the jerking of the bell-rope, and stopped short. As for the rear cars, they were brought to a standstill at once by means of the automatic brakes. I tell you they are a great institution, and indispensable to success on any road which has heavy grades to overcome."

"How do they operate?"

"Well, it is the Westinghouse patent and rather complicated. Better go out to-morrow and see the apparatus on the engine. It works somewhat in this manner: When the valve on the retort under each car is set in a particular way, the letting off of the 'straight' or ordinary air-brakes causes the compression of an exceedingly powerful spring on each set of brakes. If, however, you destroy the equilibrium by forcibly parting the air connection between that car and the locomotive, as of course occurs when a train breaks in two, the great springs are released and jam the brake-shoes hard against the wheels, gripping them with tremendous force. That's what happened instantly in this case, and those heavy cars, which otherwise would have carried their cargoes to almost certain destruction, halted in a single second."

"But — what if —" began our lady member in an alarmed tone;

ALABASTER HALL.

whereupon she was speedily interrupted by the learned gentleman who was dividing his time between dinner and lecture:

"My dear Madame, our cars, like all the rest on this admirably-guarded railway, are provided with the automatic apparatus I have described. Since it is useless to pretend to one's-self, or anybody else, that an accident will never happen, it is well to understand that every precaution known to intelligent management has been provided against any serious harm resulting when things do go wrong. In consideration of all which profound explanations I think we deserve a second glass of claret. My toast is: The Automatic Brake!"

And we all responded, "May it never be broken!"

Sleep that night was deep and refreshing. The next morning broke cool and clear, and the Photographer proposed, with nearly his first words, that we all go to the top of Veta Mountain. Only the crest of one of the spurs could be seen, and this did not appear very far away, so that those who had never climbed mountains afoot were enthusiastic on the subject. Now in the humble, but dearly experienced opinion of the present author, the old saw,—

> "Where ignorance is bliss
> 'T were the height of folly to be otherwise,"

fits no situation better than mountain climbing. I have said in the bitterness of my soul, on some cloud-splitting peak, as I tried to gulp enough air to fill a small corner of my lungs, that the man who belonged to an Alpine Club was *prima facie* a fool. Scaling mountains for some definitely profitable purpose, like finding or working a silver mine, or getting a wide view so as properly to map out the region, or for a knowledge of its fauna and flora, is disagreeable but endurable, because you are sustained by the advantages to accrue; but to toil up there for fun—bah! Yet people will go on doing it, and those who know better will follow after, and the heart of the grumbler will grow sick as he sees of how little avail are his words and the testimony of his sufferings.

It was so this time. Admonitions that upright distances were the most deceiving of all aspects of nature; that the higher you went the steeper the slope and the more insecure and toilsome the foothold; these, with other remonstrances, were totally unheeded, and three misguided mortals decided to go. Then the growler yielded—what else could he do? He had survived many a previous ascent, and could not afford to assume a cowardice that really did n't belong to him. So he chose that horn of the dilemma, and left the reader to the conclusion that in telling this tale, after the previous paragraph, he "writ himself down an ass."

All went but the Musician. Among the gentlemen were divided the photographic camera and materials, and the whisky, while the Madame set off sturdily with field-glasses over her shoulder, and a revolver strapped around her shapely waist. Dinner was ordered for two o'clock, and up we started. The Madame wrote to her friend about it as follows (the letter, I declare, smelled of camphor):

"I assure you, my dear Mrs. McAngle, that not more than a hundred feet had been gone over before the inexperienced of our party began to feel the effects of rarefied air, although thus far it was easy enough walking. There was no path, of course, and we simply tramped over a grassy slope sprinkled with flowers and covered by trees that shaded us from the sun. Gorges which were hardly perceptible as such from the valley, now proved to be uncomfortably deep gashes in the broad mountain-side, and tiny streams came down each one of them, to water dense thickets along their banks. In one place, about a thousand feet above our starting point, we came across the remains of a camp made by some man who thought he had found precious metal. Dreary enough it looked now, with its dismantled roof and wet and moldy bed of leaves.

"By this time breathing has become a conscious difficulty. I speak in the present tense, my dear, because the recollection is very vivid, and it seems almost as though I am again trudging over those sharp-edged rocks. Every ten minutes further progress becomes an utter impossibility for me, and rest absolutely necessary; but one recuperates in even less time than it takes to become exhausted, and starts on again. Nevertheless I can not go as fast as the gentlemen, who have no skirts to drag along.

"Now the comparatively easy climbing is over. Flowers and grass have grown scarce, and almost all the trees have disappeared. Nausea is beginning to annoy me, and I was never more glad in my life than now, when I discover some raspberry bushes and eagerly gather the ripe fruit, whose pleasant acid brings moisture to my parched mouth and comforts my sad stomach, for there is no water or snow here, and I know it would not be best to drink if there were.

"Even the berries are gone now. Far above and on all sides I see nothing but fragments of rocks. For centuries, wind, frost, rain and snow have been hard at work leveling the mountains. They have broken up the hard masses of yellowish white trachyte, and the dikes of black basalt into small pieces—some as minute as walnuts, but most of them much larger, with sharp-pointed edges that cut my feet. Across these vast fields the wild sheep, thinking nothing of jumping and gamboling over such steep slopes of broken stones, have made trails that cross and criss-cross everywhere. Availing ourselves of these is some help, as we all settle down to persistent, never-ending climbing.

"Up, up, up. You have forgotten how to breathe; your back and head are aching; you have found a stick, and lean more and more upon it; you look down on the back of a hawk far below you with sullen envy; you devoutly wish you had never come, but will not give up. At length a stupor creeps over you. You never expect to reach the top, but you do not care; old long-forgotten songs go through your brain and seem to try to lull you to sleep. You see in the distance one of the strong ones reach the summit and wave his hat; you are beyond sensation, and it is all a dream. Finally you stagger over the last ledge and throw yourself down on the top and feebly call for—whisky. Mrs. McAngle, I am a teetotaller; I hate whisky! But just then I would have given half my fortune had it been necessary for the one swallow which did me so much good."

Well, her companions having more strength, did n't feel *quite* so bad, though near enough so, to make their sympathies strong. The crest having been gained, the Madame lay down on a rubber coat under the cap-rock to rest, while the remainder of us dispersed in search of water. But let me quote that long letter again:

"The rocks, when I had recovered strength to look about me, I saw were crumbling lavas of two colors, light drab and dark brown. Covered, as they were, with lichens of brown, green and red, they were very pretty. At last one of the gentlemen came back, carefully carrying his hat in both hands, which he had made into a sort of bowl by pressing in the soft crown. This I soon saw contained water, but such water —foul and bad tasting, for it had been squeezed from moss. But we drank it through a 'straw,' made by rolling up a business card, and were thankful.

"Refreshed, and becoming interested in life again, the old hymn occurred to me,—

'Lo, on a narrow neck of land
'Twixt two unbounded seas I stand.'

Only the seas, in this case, were broad green valleys, and were bounded in the distance by lofty mountains, best of all Sierra Blanca, across whose peaks the clouds were winding their long garments as if to hide somewhat the sterility and ruggedness of their friends. Above them how intensely blue was the sky, and how the soft green foothills leaning against them satisfied your eyes with their graceful curves. Trailing among them, as though a long white string had been carelessly tossed down, ran the serpentine track of the railway, and the famous Dump Mountain sank into the merest foot-ridge at our feet. On the other side of the ledge we gazed out on the misty and limitless plains, past the rough jumble of the Sierra Mojada, and could trace where we had come across the valley of the Cuchara. Nearer by lay dozens of snug and verdant vales, in one of which glistened a little lake tantalizing to our still thirsty throats.

"We all had our photographs taken, with this magnificent scenery for a background—better even than the cockney-loved Niagara, we thought—and strolled about. Not far away we hit upon a prospect-hole. The miner was absent, but had left pick and shovel behind as tokens of possession. How intense must be the love of money that would induce a man to undertake such a terrible climb, and live in this utter loneliness and exposure! Yet they say that many of the best silver mines in Colorado are on the very tops of such bald peaks as this.

"At last, on asking my husband if he did not think he appeared like an Alpine tourist, I found him recovered sufficiently to say that we should all pine if we did n't have dinner soon, so we turned our faces homeward. Now I hope I have n't wasted all my adjectives, for I need the strongest of them to tell of that descent. It was frightful. Feet and knees became so sore that every movement was torture. The sun blinded and scorched me, and the fields of barren, sharp and cruel stones stretched down ahead in endless succession. Mrs. McAngle, however foolish I may be in the future in climbing *up* another mountain, I never, never will come *down*, but will cheerfully die on the summit, and leave my bones a warning to the next absurdly ambitious sight-seer. When I was on the crest, I thought what an idiot the youth in 'Excelsior' was, but now I hold him in high respect, for he had the great good sense, having reached the top, to stay there!"

Returning to Veta Pass, the promontory where the track winds cautiously around the brow of Dump Mountain—the name is given because of a resemblance in shape to the dump at the entrance of a mine tunnel—has been called Inspiration Point. I don't know who christened it; perhaps some would-be hero of a novel by G. P. R. James. If, to be in character, he "paused at this point in involuntary admiration," there was plenty of excuse, for one of the loveliest panoramas in Colorado unrolls itself at the observer's feet.

Coming up is fine enough, if you see it on such a day as the gods gave us. The Spanish Peaks, as we approached from Cuchara, were as blue as blue could be, with half-transparent, vaporous masses hovering tenderly about them; but these mists stopped short of Veta, which stood out distinct against its cloud-flecked background, majestic in full round outlines beyond the majority of mountains,—in hue purple and sunny white, with the mingling of forests and vast sterile slopes. North of it the landscape was almost hidden under rain-veils, into which the sun shot a great sheaf-full of slanting arrows of light, and beyond, range behind range were marked with phantom-like faintness of outline. A broad canopy of leaden clouds hung overhead, down from the further eaves of which was shed a wide halo radiating from the invisible sun above; and this snowy shower had stood long unchanged before our entranced eyes, making us believe that the brown cliffs, toward which we were running so swiftly, were the gates of an enchanted land.

VETA PASS.

Now, from within and far surmounting those portals, we stand gazing abroad, as in olden days they looked out of some castle tower through and beyond the great fortress arch. The typically mountain-like mass of Veta, satisfying all our ideals of how that style of elevation should look which does not abound in rugged cliff and sky-piercing pinnacles, but is smoothly and roundly huge, cuts off all northward outlook. Southward the crowded foothills of the divide between the Rio Grande and the Arkansas hide from view the central points in which they culminate,—even lofty Trinchera, whose sharp summit was so plain a landmark at the southern end of the Sangre de Cristo yesterday. Beyond these swelling domes and gables, and ridges of green and gray, were lifted the noble pyramids of the Spanish Peaks, their angles well defined in varying tints of purplish blue, and their grand old heads sustained in generous rivalry. (Illustration.) Behind us was only a piney slope, close at hand; but ahead — the world! I think no one has ever said enough of the beauty of this picture in Veta Pass. From the precipitous, wooded mountain-side where you stand, the eye follows the little creek as it glides with less and less disquietude down through the protruding bases of the diminishing foothills, into the slowly broadening valley where the willows are more dense, and the heather and small bushes have taken on brilliant colors to vie with the splendor of aspen-patches higher up; on to the hay-meadows fenced with the many-elbowed and scraggy faggots of red cedar; on past the little park where the low brown adobe houses of the Mexican rancheros look like mere pieces of flat rock fallen from the mountains; on into the midst of minute cornfields; on out, beyond the surf-like ridges breaking against the base of the range, to the blue and boundless sea of the plains.

The western side of the Pass is a tortuous descent through continuous woods and lessening hills, until you emerge upon a plain where the ragged heights of the Saguache Mountains fill the northern horizon; and as you turn southward the glorious serrated summits of the Sangre de Cristo range come into view behind and beside you, on the east. This plain is the San Luis Park, the largest of those four great interior plateaus—North Park, Middle Park, South Park and San Luis—which lie between the "Front" and the "Main" ranges of the Rockies.

It has been truly said of the Rocky Mountains that the word "range" does not express it at all. "It is a whole country, populous with mountains. It is as if an ocean of molten granite had been caught by instant petrifaction when its billows were rolling heaven-high."

Nevertheless, popular language divides the system into certain great lines. The "Front Range" extends irregularly from Long's Peak to Pike's Peak, then fades out. The "Main" or "Snowy Range," which is the continental watershed or "divide," begins at the northern boundary, in the Medicine Bow Mountains; but in the center of the State breaks out of all regularity into several branches, so that it is only

CREST OF VETA MOUNTAIN.

by ascertaining where the headwaters of the Atlantic and Pacific streams are separated, that one can tell how to trace the backbone of the continent, for many of the spurs contain peaks quite as lofty as the central chain. Thus the splendid line of the Sangre de Cristo, running southeastward, only divides the drainage of the Rio Grande from that of the Mississippi; yet the highest peak in Colorado belongs to it. The main chain, on the contrary, trends southwestward from the parallel groups in the heart of the State, only to become mixed up into half a dozen branches, all of enormous height and bulk, down in the southwestern corner. Even this is not all, for westward to Utah the whole area is filled with vast uplifts, standing in isolated groups, serving as cross-links, or lying parallel with the general north-and-south lines of great elevation. "I suppose," says Ludlow, "that to most Eastern men the discovery of what is meant by crossing the Rocky Mountains would be as great a surprise as it was to myself. Day after day, as we were traveling between Denver and Salt Lake, I kept wondering when we should get over the mountains. Four, five, six days, still we were perpetually climbing, descending or flanking them; and at nightfall of the last day, we rolled down into the Mormon city through a gorge in one of the grandest ranges in the system. Then, for the first time, after a journey of six hundred miles, could we be said to have crossed the Rocky Mountains."

Because we had ascended and descended Veta Pass, therefore, and saw on our left the seemingly insurmountable barrier which yesterday stood at our right, we had by no means got beyond the Rockies; for out there "mountain billows roll westward, their crests climbing as they go; and far on, where you might suppose the Plains began again, break on a spotless strand of everlasting snow."

VI

SAN LUIS PARK.

*The plain was grassy, wild and bare,
Wide, wild, and open to the air.*

—TENNYSON.

SAN Luis Park, exceeding in size the State of Connecticut, is identified with the earliest and most romantic history of Colorado. It was here that brave old pioneer, Colonel Zebulon Pike, established his winter quarters almost a century ago, and was captured by the Mexican forces, for at that time all this region was Spanish territory. It was here the northernmost habitations of the Mexican people, the ranches at Conejos, Del Norte, and all along between, were placed, and so became the first farms in what now is Colorado. Here were pastured the first herds and flocks of the early settlers, and the great Maxwell grant, whose ownership has been the subject of so much litigation, included a large portion of this park. To this region, long ago Governor Gilpin directed the attention of immigrants, and lauded it as the "garden of the world." Gardening is practicable here, without doubt; but colonists have found other parts of the State so much more favorable, that, in spite of its superior advertising, the park has kept nearly its pristine innocence of agriculture outside of the old Mexican estates along the principal streams.

That Colorado can ever produce cereals enough for the sustenance of a large population is doubtful. The great rarity and dryness of the atmosphere; the light rainfall, and almost instant disappearance of moisture; the large proportion of alkaline constituents in the earth, and the climate caused by great altitude, seem to handicap this region when compared with the Mississippi valley or the Pacific coast. By irrigation only, can agriculture thrive in this State; and the amount of arable land that can be cultivated without enormous expenditure for irrigating canals can hardly be considered wide enough to long supply the local population, which increases faster than the acreage under the plow is extended. The nature of the soil, and the effect of the short, hot seasons, under careful regularity of watering, combine, however, to make the product of Colorado farms extremely heavy to the acre, and of the finest quality in every article grown.

"The San Luis valley," says a recent report to the Government, "bears witness to the wealth of the produce returned by the soil under proper cultivation. In following up the Rio Grande, the Mexicans ascended divers tributary waters, and upon these and along the main river can their apologies for farms be seen. Generally content with simple existence, but little variety in the products of their land is observed The turning of the earth with oxen and a sharpened stick, the threshing by flail and trampling under foot, and the crushing of the grain between stones, can be so frequently seen, that the charm of novelty is lacking and one's curiosity is soon satiated. Progress is not their hope or desire, and, content to eke out a bare subsistence, their ambition does not extend beyond a *baile*, or the tripping of the 'light fantastic,' with surroundings that are here, as a rule, far from enchanting.

"Their cultivation of the ground tells of eastern origin and traditions, and is by irrigation from *accquias* or canals. Smaller ditches at intervals lead out from the main, and furrows of earth of varying height, connected thereto, are raised at stated points parallel to one another, cutting up the entire area into many patches nearly square and of small extent. With the planting of the seed and the main ditch filled, all the smaller outlets and various sections being simultaneously overflowed, the entire area is carefully submerged, the little furrows confining the water in each section. To the inexperienced farmer, the first successful irrigation of his land is a matter of considerable labor and pains. Besides the thorough moistening of the earth, obtained by the gradual settling of the waters, a fertilizing process is at the same time ensured. These streams carry in solution much rich and valuable material from the denudation of sections drained in their passage, which is left in deposit like a substratum of manure. The latter is never used, the farmer depending on irrigation for the supply of those constituents extracted from the soil in the growth of produce.

"The Rio Grande descends from seven thousand seven hundred and fifty feet at Del Norte, to seven thousand four hundred on leaving the State for New Mexico. Upon its western side numbers of locations are along the Piedra Pintada, which sinks a few miles from the Rio Grande, the Alamosa and La Jara, but chiefly along the Conejos, the most thickly settled of all its tributaries; upon the eastern side are the Trinchera, Culebra, Costilla, the Culebra above San Luis being on this side the seat of largest habitation. In the upper part of San Luis valley is situated the finest land of the section, with the mountain range encircling it upon the east, north and west. Exposed only upon the south, whence do not come the heavy snow storms and coldest winds, it contains the best lands for cereal and other productions. Drained by the San Luis creek, and the Saguache, its tributary, the ranchmen who have located along the streams have been rewarded for their labor by very abundant crops of all kinds. Throughout the valley large herds of cattle find ample sustenance, the property mainly of Americans, while numerous flocks of sheep of Mexican ownership, are driven to and fro. The valley of the Conejos, with its affluents, San Antonio and Los Pinos creeks, is a most fertile region. Several miles east of Conejos, during the highest stages of the rivers, in June, water from the San Antonio finds its way into the former river above the latter's mouth, forming an island. This section is especially rich, and there exists almost a natural irrigation, the Mexican ranchmen raising large crops of all kinds at the cost of but little labor therefor.

"The Alamosa and La Jara, during the lower parts of their courses upon the plains, run side by side. At the foothills they diverge, the head of the Alamosa being in the northwest, its course throughout in a generally narrow and very deep cañon, while the upper waters of the La Jara are due west. All the portions of the former that are available for agriculture, are its banks on the plain and a short part of its cañon-valley within the foothills, upon which the Mexican ranches are found. Upon the La Jara are a few more Americans than upon the former, the ranch-owners being mainly, however, of Mexican descent. A tributary is called by the geographer North Fork, but is locally known as Aguas Calientes, or Hot Springs Creek, and where its land is represented as suitable for grazing only, it is found in reality to be adapted to the agriculture of the Mexicans, ranches at intervals being passed along its banks.

"The entire course of the La Jara may be likened in its direction to a huge frying-pan in outline, the long handle upon the plain extending to the Rio Grande, the basin within the foothills to its source. Before reaching the plains the stream flows to the south, east, and north, the latter part in a steep, precipitous cañon, strewed with basaltic, which the road avoids. This road, built by the county over a natural route, is in good order, and affords the residents of the lower river easy access to its upper part, which, as we ascend and pass over the intervening rolling foothills, we find within a lovely valley, called by the Mexicans *El Valle*, to which they resort for hay. The volcanic rocks strewn along the foothills, well timbered with piñon, we leave behind us as we descend into the valley, a basin eroded from the general plateau by the waters of the stream, which has cut for itself, in its lower and more rapid descent, a small but impassable cañon. This valley, several miles long and of a varying width of from three-fourths to one and a half miles, is a beautiful spot, and has been located upon by several persons for cattle ranches. The grazing is very fine, and so nearly level is the land, that the stream, here small and at its headwaters, pursues a most tortuous course. Trout are found more abundantly than at any other point."

While we can scarcely compliment the syntax of the report above quoted, the facts are trustworthy.

Fairly out into the valley, where Ute creek, **Sangre de Cristo**, and one or two other streamlets unite to form the Trinchera, stands an old military post, Fort Garland. In one of the cañons near by was a still older one, Fort Massachusetts, now abandoned or used only as a cavalry cantonment when a larger body of troops is assembled here than there are barracks for. In 1852 and 1856, the dates when the two forts respectively were founded, the Indians—Utes, Apaches, Comanches, and Navajoes—were all troublesome, and the men were kept very busy in scouting, if not in resisting attacks. Now the crumbling buildings of adobe shelter only a score or so of men, and serve merely as a depot of supplies, a large amount of government stores being guarded here. Fort Garland is a pretty place, and from it will be likely to make his start anybody who wishes to ascend old Sierra Blanca, the loftiest peak in Colorado, whose triple head stands grandly opposite and near the railway; the United States Geological Survey is his sole predecessor.

Long ere this we had become domesticated in our cars, and now I may digress sufficiently to jot down a little description of them. As I have said, there were three, and we spoke of them as our "train." The first was a parlor-car. It usually ran in the rear, and gave us the advantage of a lookout behind, something worth having among the mountains. This car was not homelike enough to suit us, however, so we rarely occupied it, when we were stationary, except as a bedroom for our masculine guests. But when running, this car was our resort. Into it we would hustle the Madame's sewing-basket and fancy work, a lot of books and papers, spy-glasses, wraps, and luncheon, and have the gayest of times as we sped along, unconfined by limited space, unsolicitous about baggage or appearances, unannoyed by other passengers, and above all, thank heaven! safe from the peanut-boy. If we were to run at night we converted it into a sleeper. Curtains were hung up at intervals, making staterooms; easy chairs were faced, a stool placed between them, then a mattress spread across, forming a capital bed; or else we simply cleared a place on the floor, spread our mattresses down, and camped. Usually both methods were followed by different occupants. It was snug, there was good ventilation, and we slept such slumbers as seemed to prove us in the poet's category of the "just." Where a long stay was made, cots were set up, and the car became a bed-room exclusively. I doubt whether our porter enjoyed it, though, as much as we. He rarely rode upon its easy springs, and he had a constant fight with circumstances to keep it neat.

The other two sections of our train were box-cars, fresh from the shops, and of the most improved pattern. All through the trip, I may say in advance, they rode splendidly, though often attached to express trains which rattled them along at twice the speed the maker ever intended. Each of these cars had a door cut in one end, and these

SPANISH PEAKS, FROM VETA PASS.

door-way ends were placed in juxtaposition and remained so always. At first two elaborate platforms were hinged to one of the cars, bridging the space between them; but they were smashed on almost the first curve, after which we laid a series of boards down from one buffer-head to the other and took them up whenever we moved—that is, if the porter did n't forget it, or get left. Here comes in the chronicle of our steps, the portable stairway by which we ascended and descended to and from our elevated house ; *sed revocare gradus*,—" but to recall those steps" in their entirety would, I fear, be a hopeless task. The first set we had fell under the wheels and immediately became of no further interest to us. Then our invaluable forager found this second set, and thereby saddled himself with a responsibility he never could shake off. The whole Denver and Rio Grande railway corporation seemed to be bent on their destruction. Time fails me to tell of the numberless occasions when they were apparently crushed by some jar of the cars, as they stood in position at a station, and of the wrenchings that required a new hammering and more spikes to correct. But watched jealously by the porter, and lashed securely on the end of the car when we moved, they survived it all, and gave us *facilis decensus* from first to last.

One of these box-cars became kitchen and commissary office. A partition was thrown across one-third of the distance from the end, forming a room for our porter and also a place of storage for our supplies. There was everything in there, from a pepper-box to a mattress, and from a lamp-chimney to a Winchester rifle. It had a table which might have been let down, two windows, and sundry racks and clothes-hooks. The remaining two-thirds of the car was devoted to the kitchen. One corner contained a monstrous ice-chest, and opposite it stood a huge wood-and-coal box, which it was the constant ambition of our boy to keep piled with kindling stuff almost to the ceiling; the result being, that frequently his improvised racks would come to pieces with the jarring of some rapid run, and the fuel be heaped up " mighty promiscuous" on the floor. The other corners of the kitchen held a fair-sized cooking-stove, securely bolted, and, lastly, an iron water-tank, as large as a barrel and mounted on a stand. With this water-tank we had a long contest. The face of our first colored cook, never much more cheerful than the big end of a coffin, took on a doubly rueful aspect at the conclusion of our first day out. The tank had been well filled before starting, but the cover fitted so loosely that half a barrel or so of the liquid splashed out, and the floor of the car was like a little sea. The Photographer generously sacrificed a blanket to spread across underneath the cover, and we were careful afterward not to fill the tank quite to the top; but it always shot jets and sprays down the back of your neck when you least expected, if you went near it when in motion. Then one day the faucet burst, and deluged the place with a stream like that from a hose-pipe. Next it fell to leaking, and so to the end of the trip we had that persistently mischievous tank to contend with. Beside the stove stood a narrow cupboard, the top of which was intended to be the kitchen-table; but we found water leaking through into the flour, etc., underneath, and so built another table, hinging it to the opposite side of the car, between the tank and the fuel-box. There were plenty of shelves and racks; and, the two side-doors having been fastened shut, the walls of the car were soon garnished with all sorts of wares that could be hung up. After a week it was learned how to stow everything so well that almost no breakage occurred.

The dining-car was exactly similar in size, twenty-four feet long by seven feet wide. It had four windows, and we used to slide back the great doors on one or both sides when the weather was warm and pleasant. If cool or stormy we locked them, wedged them tight and caulked the cracks, yet could never quite keep out the gales. The wind, I found, bloweth not only where "*it* listeth," but also where *I* listed. We thought it a very cheerful place, as we entered this snug home—for it was the "living-room" of the train—after a hard tramp, or gathered about the dinner table in the strong rays of mail lamps, and the softer light from railway candles. The gayly striped *portière* shutting off the Madame's little nook of a bed-room at the rear end of the car; the bright oilcloth that covered the floor; the rich oak-brown of the paint on the door-frames, wainscoting, and stanchions that at frequent intervals supported the roof; the ruddy glow of the Turkey-red cloth filling all the panels, and the pictures, books, Indian pottery, burnished firearms and bits of decoration here and there, made a picture that never lost its cheer and air of comfort. Here were my friendly books and writing-desk, with all the little literary appliances, and pigeon-holes full of manuscript, memoranda and correspondence. Here was the easy chair behind the spindle-shaped, upright stove. Here was the Madame's rocking-chair and her work-stand, while the parted curtains let us peep into a diminutive but carefully convenient boudoir just behind her. Here stood her wardrobe—a trunk which lost its identity under the warm zigzags of a Navajo blanket; and here our hospitable dining-table, around which, perched on camp-stools, we ate good food with royal appetites, drank red wine with keen delight, and summoned all the imps of fun to laugh with us over quips and quirks to which, no doubt, the mad spirit of the day lent more wit than the brains of their makers. Shakespeare says,—

> "A jest's prosperity lies in the ear
> Of him that hears it, never in the tongue
> Of him that makes it."

Here work was done, too. Have I not seen the Madame busily sewing, and quiet? Did not the Artist often paint, I know not how long, without speaking?—I know not how long because I was so intent upon shaping this chronicle you read. If our trip had been all picnic and void of serious purpose, we should not have enjoyed it half so well. Charles Lamb asked pettishly,—

> "Who first invented work, and bound the free
> And holiday-rejoicing spirit down?"

But surely our holiday was fraught with a deeper zest, because our not too onerous duties now and then encroached upon our pleasures, and so made us value merry times the higher.

Well, now you may understand how and where we lived, and moved, and had our work and play. It was a warm, snug, handsome home and office, bed-chamber and kitchen on wheels. There were little hardships and annoyances, no doubt, but why remember them? *Le diable est mort!*

The railway down San Luis Park is straight as a surveyor's line, and trains

SANGRE DE CRISTO SUMMITS.

often run at a high rate of speed with perfect safety. At Alamosa it halted in construction for a long time. The town then became the forwarding point for all southern Colorado and New Mexico. Very large commission houses were placed there, enormous trains of wagons and pack mules were coming and going, stages left daily for Lake City and Gunnison, Saguache and Pitkin, Tierra Amarilla and the lower San Juan, Taos and Santa Fe, and the vim and excitement of an outfitting station prevailed. But presently the railway moved southward and westward, and Alamosa settled down into a quiet yet prosperous place, with a local agricultural population to back it, and the headquarters of the second division of the railway which extends to—to—well *towards* Mexico.

Twenty-nine miles south of Alamosa is Antonito, where the line branches off to the San Juan country. The town is supported by the money the railway and the passengers spend, and is quite uninteresting; but over to the westward is the larger and older village of Conejos, which is better, though "distance lends enchantment." Conejos means *hares:* probably the Mexican pioneers found a superfluity of jack-rabbits there. The place has been a farming and grazing center of supplies for many years. Along Conejos creek are numerous small Mexican ranches, good enough types of their sort (we shall find far better ahead), but the town itself has been Americanized until its claim to being a Mexican plaza is about lost; nor have the innovations added to its interest in any degree. In a real Mexican town, for example, the church is always an entertaining place to visit, because it is ruinously ancient and strange; but here the large, well-conditioned structure has been roofed, painted and modernized until it is not worth a glance except from the point of comfort and security from decay. Annexed to it is an academy for boys, and another for girls, both under the charge of priests and nuns of the Roman Catholic Church. These schools have no counterparts among the Mexicans nearer than Santa Fe, and have a wide reputation.

Lacking interest for the tourist, the practical man will learn that Conejos is a very fair business place in certain lines. It is the headquarters of the sheep and cattle men of the San Luis Park. In sheep, I learn that although about two hundred and fifty thousand are sold out of the park annually, fully five hundred thousand are left. The large majority of these are of the inferior sort called Mexican sheep, which are worth from one dollar to one dollar and twenty-five cents a head. The better minority sell at one dollar and fifty cents and two dollars a head, and this minority is increasing through a constant effort to improve the breed by introducing highly-bred Merino and Cotswold rams. The average yield is two and a half pounds of wool annually, and the product is shipped almost entirely to Philadelphia, for use in making carpets. Cattle is less an industry here, because the sheep are so numerous as to consume most of the pasturage. Something like ten thousand head, however, are able to exist in the park and adjacent foothills, and are sold to great advantage.

Nearly midway between Alamosa and Antonito, and easterly, but within sight of the railway, the Mormon settlements of Manassa and Ephraim have been founded, and have now a population of about six hundred. These people do not practice polygamy, and are frugal, industrious and prosperous. They are under the jurisdiction of the church in Utah, which also maintains similar colonies in the corners of Arizona and New Mexico adjacent to the Utah and Colorado line.

VII

THE INVASION OF NEW MEXICO

> There are mists like vapor of incense burning,
>   That are rolling away under skies that are fair;
> There are brown-faced sunflowers dreamily turning,
>   Shaking their yellow hair.
>
> —Mrs. C. L. Whiton.

OUR stay was comparatively not as long as our talk in sandy San Luis, for we soon left its pastures behind and were steaming southward, but with slower and slower speed. Again we were twisting our toilsome way up the valley's "purple rim," since it was easier to go over the high bank than down through the rugged cañon, where the wagon-road runs. The summit of this ridge, beyond which lies the valley of the Rio Grande in New Mexico, is not attained until you reach Barranca ("a high river-bank"), sixty-five miles from Antonito, and one of the most attractive spots in the region. The altitude is over eight thousand feet, and the air of that perfect purity extolled in all that is written of mountain districts. The station, with its half dozen accompanying houses, all owned and occupied by those in the service of the railway, stands in the midst of acres of sunflowers, which all summer long spread their yellow disks to the full gaze of the sun, and dare him to outshine them. I have seen sunflowers before, but never in such masses and splendor of tone as here. Near by one catches sight of the green of the leaves and stems between, like the mottled plumage of some canaries, while the mass of chrome-yellow atop is picked off with maroon dots of seed-centers. Distance loses these details to one's eye, and gives only a billowy stretch—a glorious sulphur sea, intense as burnished gold, rolling between you and a dark green shore of piñon foliage. This August landscape, indeed, divides into three great portions, relieved by few variations, yet never for a moment monotonous. In the foreground are the brilliant, owl-eyed flowers; above them the stratum of well-rounded tree-tops, blackish in shadow; after that the far-away mountains, delicately green, or deep blue, or washed with an amethystine tint. Arching over all bends the cloudless azure of the canopy.

Our cars safely side-tracked, the Madame and I wander aimlessly about during the warm afternoon, while the Musician takes his rifle and saunters away down the tapering track, and Photographer and Artist set resolutely off for a tramp to the top of some buttes. Returning in good time for our twilight dinner, the rifleman brings no game, but reports having seen a cotton-tailed hare and some ravens. The climbers come in very weary. Their buttes were far away and lofty. From their summits they could distinguish Fernandez de Taos, and are not surprised to learn that more freight is gradually being transferred thence from here than from Embudo, the designated station for Taos. They could see more grand peaks than they could count on their fingers and toes; and told us about the road to Ojo Caliente, the Warm Springs of New Mexico, famous for almost three centuries.

This was an objective point twelve miles away; but the stages which now run, had not then begun their daily mail-trips, so we had to dispatch a messenger for a vehicle. Three Mexicans lay stretched out on the station-platform asleep. They had lain there all day waiting in lazy patience the coming of the pay-car which owed them a few dollars. For *uno peso*—one dollar—one of them consented to take a message to Antonio Joseph, who owns the springs; and, mounting his burro, scampered off with an appearance of tremendous haste, which doubtless diminished as soon as he had placed the first thicket behind him.

The Madame and I took our chairs into the shadow of the diningcar, and read and sewed while the sun sank reluctantly down. It was very quiet, the humming of wild bees and wasps furnishing almost the only sound. Soon I noticed that between our glances a mound of earth had been thrown up about a dozen feet from where we sat. A moment later there was a stirring in the nearest clump of sunflowers, some soil was tossed up, immediately followed by a brown nose and shaggy little head, which instantly disappeared, only to come up again, pushing before it a handful of dirt from its tunnel. This was repeated a dozen times or more, at the end of which the little workman came on top and surveyed his surroundings. He saw us, but as we kept still he took no alarm, and presently let himself down backwards into his burrow. He was not gone "for good," however. In a moment the blunt, stiff-whiskered snout and black eyes peered out, and made a grasp at a stem of the sunflower. It was large and tough, and the first bite only made it tremble; but a second nipped it off. Then seizing it by the butt,—he was wise enough not to drag it leaves foremost,—he pulled it down into his hole, and apparently carried it to the innermost chamber, for some minutes elapsed before he returned for a second flower stem, to add to his winter stock.

We knew him well enough. He was the gopher, a large kind of ground squirrel, not easily to be distinguished from a prairie-dog in race, color, or previous condition of servitude. If any one desires to know more about him, let him "look up the authorities," as Professor Polycarp P. Pillicamp would say.

That evening we sat in the parlor-car, or rather lay easily on the

SIERRA BLANCA.

couch-like chairs reclined to their utmost limit, and listened to the Musician's violin, while the glorious banners of the retreating day paled slowly from view. Opera, sonata, opera-bouffe, ballad, came tenderly to our ears and floated away out over the sun-flowers and into the piñons, where the few birds awoke to listen; and perhaps were wafted out to the rolling plain, whence at intervals came faintly in reply the long, yelping howl of a coyote.

In the morning we were awakened with the announcement that our conveyance was ready. It had come at midnight, and proved to be only an ordinary farm-wagon, with springs under the seats. The road wound through the clustered trees and crossed open glades, as though in a nobleman's park; passed the rocky buttes, whose jutting cliffs were strangely picturesque in their Grecian morning-robe of angular shadows, and descended a long, stony hill to the mesa, which thereafter it traversed for ten miles. This mesa was a great table-land of sand. Whence it had drifted was widely discussed; but the more I think about it, the less I believe it had drifted at all, and the more I incline to the opinion that it *remained*, while a great amount of formerly overlying earth and partial rock, unprotected by the thick capping of lava shielding the larger part of the surface, had been swept away by water. There were plenty of gauges to help to a judgment upon the extent of this denudation. To the right, a long mesa, with steep sides, extended out like a promontory, whose level basaltic surface was a thousand feet above us; yet that was a valley long ago, for the lava took it as a channel. Behind, across the Chama, the Jemez mountains lifted themselves in rugged outline against the southern sky. They are volcanic; but, almost as high as they, towered the flat-topped, butte-like mountain of Abiquiu, which is not volcanic, but of sandstone, and stands as a mark of the former level of the country there, on the day when the hot basalt came hissing down from its fiery spout now lost to our tracing. On this dry, sandy, cactus-loving upland grew the rich grama-grass, another name for it being panic-grass. It has a seed-head which is neither panicle nor spike, but a perfect little one-sided brown feather, about an inch long, hanging stiffly at right angles to the stalk, and at its very summit. It is not only odd, but very beautiful, for the grass grows thinly, and every plumelet is visible.

Our progress was slow and monotonous; we had exhausted our conversation and were getting tired of everything, when the driver pointed out a red hill to the right as our destination, and presently, descending a steep bench, we turned up the valley of Ojo Caliente, whose former banks, like those of all these southern streams, were from one to five miles apart, and very precipitous. Between these the river wound its way, crossing from side to side of the valley, or pursuing the safe "middle course." In the bends of the stream were Mexican farms and the most dilapidated of adobe houses, some part of nearly every one of which had so fallen to pieces as to be uninhabitable. Our guide said the land was poor, and we believed him. Everything showed that the people were poverty-stricken and almost in barbarism, yet they had an abundance of land, pretty well watered, and great flocks of sheep and goats; we met a single flock which probably contained fifteen hundred sheep. Some of the dwellings had wooden gratings in place of windows; and the doors, made with auger and axe, had been rudely carved in an attempt at decoration, which time, smoothing and tinting, had rendered very attractive to our unaccustomed and curious eyes. Behind the best houses often would lie an unfenced bit of old orchard, grown almost wild. Such a half-ruined plazita, with its carved doors and grated windows; with its corner of goat-corral, and conical ovens at the side; its grassy roof, and high, gnarled trees overhead; its background of river-bend and cornfield and red rocks and distant misty mountains· most of all, with its foreign humanity peering out to see who was passing, made a picture which threw our art-devotees into ecstacy; and as each was passed they declared they would sketch it—when they came back! That the declaration was kept you have evidence, though modified into a general view of Ojo Caliente.

Four miles up from the bend the springs were reached, and we gladly sat down to a dinner beginning with Baltimore oysters. These springs are hot, but endurably so, after one has tempered up to it. They flow from under the cliff on the eastern bank, and are thence led into the bath-houses close by. Excepting the hotel, which will accommodate from fifty to seventy-five guests, the only other building is a large supply-store; but you will usually find a great many people living in tents near by. These warm springs are noted for their curative and healing qualities, and have been visited for many years by invalids, with miraculous results. They do tell some wonderful stories of relief given to rheumatic and paralytic patients; while diseases of the skin *vamos* at once, as a Mexican attendant phrased it. Such an effect is to be expected, when you find heated water analyzing into the following constituency:

Sodium Carbonate	196.95
Lithium "	.21
Calcium "	4.17
Magnesium "	2.18
Iron "	10.12
Potassium Sulphate	5.17
Sodium Chloride	33.03
Silicic Acid	2.10
Total	272.52

The fact that the latitude (36° 20') and inland situation give a mild, equable climate in winter, and the altitude (6,000 feet) makes the summer air sweet and invigorating, should be taken into account, however,

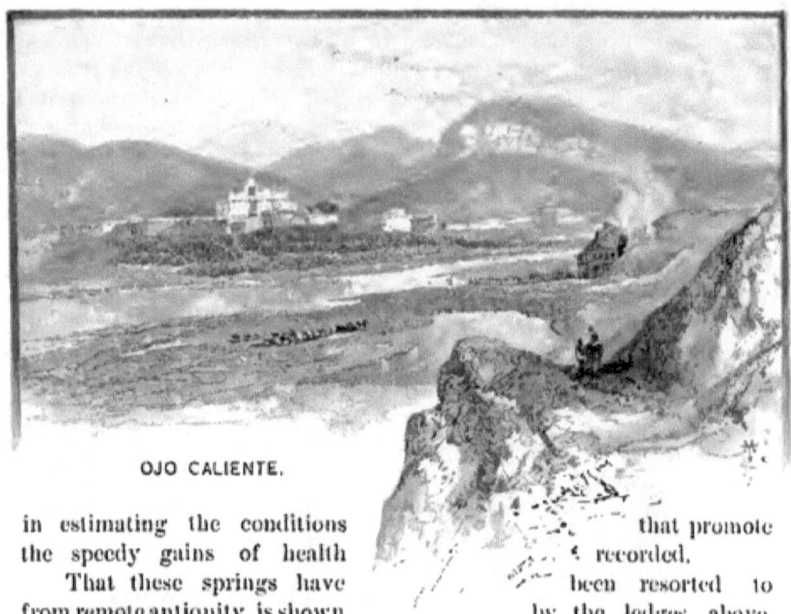

OJO CALIENTE.

in estimating the conditions that promote the speedy gains of health recorded.

That these springs have been resorted to from remote antiquity, is shown by the ledges above, which are covered with very ancient, almost obliterated, ruins of those cliff-dwelling aborigines whose houses and pueblos are scattered in such profusion over the cañons tributary to the Rio Colorado and the lower Rio San Juan. We heard that many skeletons and relics had been found there by casual excavating, and so went up to try our luck. We could trace not only the bounds of several closely grouped pueblos, but in many cases even the estufas and the straight walls of the separate rooms. A little shoveling at once showed us that these were made outwardly of uncut stone, and inwardly of adobe, which resisted the pick, while the loose earth within was easily removed. We could only "coyote round," as a western man calls desultory digging, but saw how rich a treasure to the archæologist would be exposed by systematic excavations. In searching for the stone *metates* and *las manas*, which then as now constituted the corn-crushing apparatus of the common people, the Mexican peasants have disclosed many ancestral bones, and we kicked about parts of human skeletons lying bleached, on the surface, at half a dozen places. At last, by chance, we struck a skeleton ourselves. It was that of a young person, for the wisdom teeth had not yet risen above their bone sockets, and the sutures of the skull were open. The bones were disordered, so that we obtained only a few, and the head had been crushed in. The same rude dismemberment and lack of burial is said to characterize all the skeletons discovered, and they are always found within the walls of the houses. The local theory is, that an earthquake overtook the town; but I believe that the pueblo was attacked and captured by enemies during the wars which we know finally resulted in the village-people being driven out of all this region, and that it was burned over the heads of the citizens, many of whom were killed within their very homes. The presence of charcoal all through the mounds of ruins, with various other circumstances, confirms this reasonable explanation.

We noticed fragments of pottery scattered everywhere. Some whole jars have been exhumed, I was told. Such ancient ware, uninjured, would be of priceless value, but probably it all fell into unappreciative hands, who despised its rudeness in comparison with the smoother modern ware. The samples we secured showed a close similarity to all the broken pottery strewn about the ancient and impressive ruins in the Mancos and other cañons of the San Juan valley, and, like them, had preserved their colors in the most wonderfully brilliant way. Flakes of obsidian (volcanic glass, which the settlers usually call topaz, or Mexican topaz) were very common, and I picked up one large core, whence scales had been chipped. They used this excellent material for their arrow-points and spear-heads, and we bought and were given a score or more of very fine specimens of such obsidian points, but found none except some broken ones, during our hurried look. We were told that a javelin-head of this material, over a foot in length and exquisitely worked, had been dug up here by a fortunate prospector for relics, and that he had refused fifty dollars for it.

Opposite the hotel and springs was a poor little Mexican hamlet called also Ojo Caliente, where an odd old church invited inspection. But between us and it

"There's one wide river to cross,"

—and the bridge gone. What then? The Artist, the Photographer, the Musician, "all with one accord began to make excuse." It was left for the only remaining male member of the party to make the effort, nor did he propose to wade; but how? The whole circle shrugged their contented shoulders and answered, "*Quien sabe!*"

Down in front of the hotel stood a cross-eyed Mexican with a vicious-looking black burro. Yes, he would let the Señor Americano take him, but he could not go with the Señor, because of the rheumatism in his knees, for which he had come over to the waters. So the "Señor" marched down to the post to which the burro was connected by a small rope looped about his neck. The untying of that rope was the scene for an action, Señor vs. donkey. The sarcastic remark of the Musician, "Now you have met your match!" was scarcely heard. It was not the Señor's vocation to chase that black burro around the yard, but he made it so without hesitation for a few minutes, devoting himself with the utmost diligence to the duty. The extreme levity of the idle spectators showed how utterly unable they were to appreciate a really good piece of burro-chasing when they saw it. Finally the course of the work brought the operators in close proximity to an old locust tree that had not cumbered the ground in vain with its useless trunk, as it had seemed to do for years past. The Señor skillfully put the donkey on the other side, and dexterously wound his end of the line around the sturdy trunk, whereupon the burro, like grandfather's clock, "stopped short." So would the adventure have done, had not the Mexican brought his squint to bear upon the scene, and, after a calculating survey, hobbled rheumatically to the Señor's assistance. Clasping both arms enthusiastically about the donkey's thick neck, he made signs for the cable to be cast off and the Señor to mount.

The saddle consisted of a pair of wishbone-shaped wooden crotches, fastened together on each side by a cross-bar at their lower extremities. The whole was then covered with raw-hide, which by its shrinking made the affair solid, while a cinch of the same material secured it to the little beast's back. A sheepskin was spread underneath, in lieu of a blanket, and wooden stirrups dangled by rude straps at the sides. It was a matter of agility to get into this primitive saddle, and the stay was likely to prove extremely brief, for the moment the Mexican let go his loving embrace, the burro ducked his head and made off in a swift, short circle, which came near disposing of the Señor at a tangent, through centrifugal force. Resisting this philosophical demonstration by locking his legs together around the burro's body, he finally overcame the circular intention by pounding the brute's head on one side, for there was no bridle and bit with which to guide him. The lookers on averred afterwards that it was as good as watching a yacht turn the lightship, to see the rolling skill with which the Señor veered away toward the gate, stumbled across the stony bottom, and dashed into the swift river. He himself remembers the devout thankfulness with which he found himself unwet on the other side, and the terror with which he discovered that his animal had broken into a gallop that threatened to dislocate every rib and rattle down his vertebræ, as a child tumbles over a pile of letter-blocks. What could he do? If it seemed almost impossible to stay on, it was altogether so to get off. There was no halter on which to pull, no mane to grasp, and frenzied *whoas* only urged that wicked donkey faster. But a happy thought came. He had heard a fruit-seller at Conejos say *chee! chee!* to his burros. Whether they stopped or went faster, after it, he couldn't remember, but it was worth trying. *Chee! chee! chee!* burst from his frantic lips. Instantly the beast came to a standstill, almost impaling his rider on the sharp pommel. It was a success, and his anatomy was safe again. After that, control was easier. A dig of the heel in his ribs made the burro go; a bang on the side of his head steered him away from the wrong direction, and a blow on the other side taught him he had diverged too far from the middle course, while *chee! chee!* stopped him altogether. So with trepidation and shying in a corn-field, and perilous climbing of steep rocks, at last the hamlet was reached, and the labor of dismounting painfully accomplished.

In the door of one of the low mud houses sat a woman, nearly hidden under the usual black shawl, which she had now drawn down over

EMBUDO, RIO GRANDE VALLEY.

her swarthy face. The Señor advanced and doffed his hat. You are a Spanish scholar, yet perhaps would not have understood as well as that peasant woman, had you seen or heard the conversation.

"Waynass decass, Seenyora," began the tourist.

"Buenas dias," came faintly out of a fold in the mantilla.

"Yocayrolaverolaeglahssay," was the Señor's next parrot-like remark, evidently understood by his veiled listener, for, pointing to a little man slouching past, she answered:

"No tengo llave—allí!" and disappeared in the cave-like darkness of her windowless dwelling. Meanwhile the man had gone on, sublimely indifferent to the Señor's cries and beckoning, and when followed, was found in the midst of his half-naked family, greedily devouring a melon, which he had opened by dashing it to pieces on the stone door-sill, and was now gouging out with his knuckle. After he had quite finished this pleasing operation, he got the keys of the church, and, accompanied by a little girl, led the way to the sacred edifice, whose outer court, surrounded by a mud wall a dozen feet high, was secured with a padlock.

The church itself of course was built of adobe, the façade being supported on the right of the door by a great sloping buttress, which was not only a brace, but had served in place of a ladder to those who built the roof and parapets. At each corner, in front, a little protuberance hinted that the architect had side-towers in his mind, while the center was carried up into a low gable, surmounted by a square bit of clay work and timber, bearing a wooden cross and sustaining a home-made bell, whose greenish and rough-cast exterior gave it an appearance of the most corroded antiquity. Recent rains had evidently damaged the walls very much, for great hollows had been washed in them.

Unlocking the axe-hewn and wooden-pinned doors, always innocent of paint, the Señor and the Mexican uncovered their heads, and the little girl at once knelt down, crossing her hands on her breast. Unlike the old sister who exhibits the ancient chapel of Our Lady of Guadalupe at Santa Fe, and never leaves her knees during the whole visit, however, this pious young maiden sprang up in a minute and trotted round, as full of curiosity for the white stranger as he was for *la yglesia*.

This poor church was more forlorn than most of its fellows. The clay floor had lately been a pool of water, and its drainage had ploughed deep furrows and left soft holes. The little round box of a pulpit, painted in streaks of red and blue, had replaced its lost stairway with a ladder, and its sounding-board was a spoon-shaped piece of plank about the size of a chair-seat, inside which was traced a white dove on a blue ground, its wings outspread in full conventionality. Nothing so good as a draw-shave had ever worked out the supports of the altar-rail, behind which the floor was planked. The altar itself bore in the center an image of the Virgin Mary, about half life size, dressed much like a great doll. On each side of her were tall tallow candles, set in rough holders whittled out of billets of wood into a rounded pillar form; and all about the altar were small sconces stamped out of tin (generally devoid of mirror), and cheap prints, colored and uncolored, of the Savior wearing the crown of thorns, Madonnas, and other sacred subjects.

The altar-cloth was calico, trimmed with frills and flounces of cotton lace and red muslin, more or less ragged and dirty. On either side of the altar, facing each other, hung crosses bearing wooden figures of Christ crucified. These also were about half life size, and were naked, except that one had a piece of cotton twisted about the loins, and the other had a short skirt of dirty tarletan, suggesting the ballet. These effigies were painted a dull white, and hung in the most agonizing attitudes,— suffering intensified by the long-drawn lines of the haggard faces, the slant of the eyes, and the dropping of the lower jaw. To produce a more horrible representation still, the carver had given the forms extreme emaciation, the ribs standing apart, the abdomen sunken, the bones and cords of all the limbs dreadfully prominent. Add to this cadaverous appearance a network of red streaks tracing the principal veins, and great splashes and runlets of blood, and you have an image awful beyond conception. Besides these large models, there was a little one of the same style, which I should have been tempted to have sacrilegiously stolen, had not the keeper been watching me closely; and in several niches, small, tinsel-clothed puppets, which the man told me were San Francisco, Patron of the Church, and our Lady of Guadalupe, who heads the list of sanctified virgins in all the Mexican churches. Standing in these little holes in the half-whitewashed wall of mud, under their ragged little curtains, the corporal's guard of saints looked very forlorn; and I do not wonder the peasants refuse to go into the building after dark, no matter how fast they may mumble their prayers.

More interesting than the images were some silken and fringed banners, decayed almost to shreds, and the spear-points of their staves well-rusted, which once belonged to the Spanish soldiery; for this church is one of the oldest in the new world. Centuries have rolled over its adobe walls, and its roof of closely-set logs and adze-carved brackets, has echoed to the clank of men in armor, as well as to the chant of half-Indian farmers and shepherds. It is rude and ugly and barbaric, representing a phase of Christianity in some respects far worse than the simple religion those Indians over at the Pueblos thought good, a thousand years ago. But the little church is not to be despised, and the awe-struck faith of its miracle-loving parishioners may be more acceptable than the gilded worship of many a rich and learned congregation nearer the sea.

VIII

EL MEXICANO Y EL PUEBLOANO.

THEN they descended and passed through the luxuriant yellow plains, the sunset blazing on the rows of willows and on the square farm-houses with their gaudy picture over the arched gateway, while always in the background rose the dark masses of the mountains, solemn and distant, beyond the golden glow of the fields.—WILLIAM BLACK.

OME just in time from Ojo Caliente, we hooked our cars the same evening to the never-tiring express, and trusted ourselves to its guidance without a thought of danger. When daylight had fully come, and from the "purple-blazoned gateway of the morn" the sun was begging entrance at our curtained windows, somebody—I think it was the Photographer, a man utterly without nervousness or regard for it in others — startled all our tranquil slumbers by the shout, "*Comanche!*"

It was not Indians though—only a respectable sort of cañon, with great black walls, and rugged hills wedged apart by the stream, and the train hanging invisibly half-way betwixt top and bottom, always going in and out of nooks and gulches, always gliding down nearer the water, until finally, between strange farm-fields, the noble Rio Grande came in view, and once more we ran upon a level track. Emerging from Comanche Cañon, a bend to the southward is made along the western bank of the lower part of the cañon of the Rio Grande. In many portions of this narrow valley, only about twenty miles in length, features of great interest to the eye occur, equaling the walls of Comanche, which was itself ignored until the railway brought it to light. The river here is about sixty yards wide, and pours with a swift current troubled by innumerable fallen rocks. To-day it is swollen and yellow with the drift of late rains, but in clear weather its waters are bright and blue, for it has not yet soiled its color with the fine silt which will thicken it between Texas and Mexico.

On the opposite bank, near the level of the river, runs the wagon road that General Edward Hatch, formerly commander of the department of New Mexico, cut some years ago to give ready communication between his headquarters at Santa Fe and the posts in the northern part of the Territory and in southern Colorado. This is the track now followed by all teamsters, but the old road from the south to Taos ran over the hills far to the eastward, passing through Picuris.

An odd conical hill (shown in our engraving) stands near the mouth of the cañon, dividing the current of the river. Noticing its resemblance to a funnel, the Mexicans called it Embudo, and the adjacent station takes the same name. Embudo is chiefly important as the point of departure for Taos, thirty miles distant.

While breakfast was preparing we were interrupted by the sudden apparition at the side-door of our car of two long ears, then a forehead, bulging by reason of the bushy hair that covered it, and immediately afterward the neck and shoulders of a donkey. But if you say *donkey* down here few comprehend you. The proper word is *burro* (boó-ro). This animal bore upon his round back a small saw-buck saddle, from each side of which hung a square panier of wicker-work. These paniers were not nailed, but the willow sticks of which they were made were bound into place by thongs of rawhide. On top, between them, was lashed a third square basket, which would hold a half-bushel. Though this seemed very bulky, it really was a light load for the little beast, and he stepped along briskly ahead of the wrinkled old Mexican who owned him. Shining through the wicker receptacles we saw green rinds, and sang out,—

"Melones?"

"Si, Señor," came the husky answer, whereupon the burro was seized by the tail and brought very willingly to anchor. Slipping several of the sticks out of their leather-loops, half a dozen long yellow specimens, something between a melon and a cantaloupe, were held up for our inspection. We hammered them with our knuckles, testing their soundness, and finding some to suit, enquired the cost,—

"Cuanto pide vm. por estos melones?"

"Dos realles!" (two shillings) was the reply; so we bought three at an outlay of seventy-five cents.

They proved muskmelonish and somewhat tough, but by no means bad. There seems to be no reason why much better melons should not be raised, since the conditions are favorable and every farmer does more or less at it. This question *why* served to spice our chat at luncheon. It was ultimately concluded that the continued degeneracy was due to the fact that all the good ones were stolen and eaten, only the very poorest being left to mature their seeds; thus the worst, instead of the best, were used to propagate from. I recite this, to show the thoughtful reader that we are not always frivolous, but often introduce grave themes into our discourse, and discuss them in a philosophic way.

Attaching ourselves to the locomotive of a working train, after the noon repast, we were hauled down the valley three miles, and given an opportunity to watch the men repair track that had been lately torn to pieces by water, two or three culverts having been swept out and the road-bed completely uprooted. The hills at that point slanted down to the river in a long treeless sweep, sown so thickly with boulders of basalt, from the size of a bushel to that of a barrel, that even the sagebrush could find scanty footing between. Down this long slope, from the mountains behind, had come one of those raging precipitations of unmeasured rain to which the West has given the expressive name "cloudburst." Truly, when one of these incidents of Rocky Mountain meteorology occurs, "the windows of heaven are opened." To such a torrent the natural rip-rap opposed a very slight obstacle. The heavy and closely packed rocks were lifted and rolled and hurled headlong as though they had been a child's marbles. Wherever any earth or mere gravel was met, it was plowed up and dashed away in a moment, while as for the railway bed, its embankments were demolished, its cuttings filled, and such heaps of stones piled upon its distorted track in some places that no attempt was made to dig it out, but new rails were laid in a different spot. They were rough and irregular enough, but we went safely over. Against these cloudbursts no railway in this region can provide; and there is nothing to be done but rebuild as quickly as possible. The skill, energy and marvelous speed with which the section men do this, and the character of the temporary track over which they run their trains until a better one can be constructed, excite the surprise of every one. Railroading in the West is as unlike the similar pursuit in the Atlantic States as a Colorado silver shaft is a contrast to a commonplace granite quarry.

We had observed on the further side of the river, where the flat lands were continually widening between the stream and the hills, signs of Mexican habitancy, and at the washout discovered a chapel of the Society of the Penitentes, into which the flood had broken a great gap near the foundation. It was a rude little house of mud, but well plastered within, and perhaps had been intended as a dwelling in former

NEW MEXICAN LIFE.

days. Creeping in through the breach, we found no furniture, but a pile of a dozen or more wooden crosses, which had been carried there by the doers of penance at Easter. The smallest of these crosses was more than ten feet in height, and its beams at least six inches in diameter. As to the heaviest, I doubt if I could have lifted it fairly from the ground. Yet the poor sinners had managed to get them across their shoulders, and so had dragged them hither, with many pains of outward penance and fearful flagellations of conscience, but with rich reward of pride before earthly eyes, and promises of glory in the world to come. From where they had been brought, or by whom, there was no record; but their ends were worn diagonally to a sharp wedge by long scraping over the stony soil. In addition to these were several small crosses of lath, which had been borne by the priests, typically; some tin and wooden candle-holders, curious little lanterns, and one of those rude religious portraits on woods, which are so common throughout this section, and which are preserved reverently among the Mexicans for generations.

The Penitentes are a sect within the Church, which the priests are said to have been discouraging. Perhaps this has had some effect, for the custom is in decay, a result due more to the railway than to the cathedral, I fancy. During the greater part of the year the Penitentes sin and are sinned against like other people, but in the spring they atone for it by wearing coarse clothes under a sort of sacrificial robe, and by torturing and starving themselves nearly to death. Walking in processions, masked beyond recognition, enduring constant castigations from each other, bearing over the roughest roads and across country the heavy crosses we have seen, and with the "pride that apes humility" enduring the utmost suffering, they consider themselves to have laid in a stock of grace sufficient to over-balance all possible crime during the coming twelvemonth. The practice has a long history, but amounts to an American survival of the Flagellants of Europe.

A few miles below, the Mexican farms and orchards became more frequent, the little settlement of Joya was noticed, Plaza Alcalde passed by, and the wide, fertile plain of San Juan opened to our view. Skirting the western edge of this (for the river keeps close under the high bluffs on that side), we ran five or six miles, until a triangular parting in the bank opened to the westward, where we halted on a side-track near the old adobe village, but new railway station, of Chamita. The Rio Chama flows into the Rio Grande here, and a broad valley area is the result. The whole of this, which is easily irrigated, is under tillage, and just now looking its best. It is therefore a green and prosperous landscape we gaze upon, bounded by reddish benches which the setting sun brightens into splendor, and shut in by blue, lofty, cloud-capped hills, beyond which stand the guardian summits of snowy ranges.

Up the Rio Chama cultivation extends almost uninterruptedly for many miles, and there are several villages or *plazas*. Chamita itself is on this side—a cluster of scattered houses along the bluff through which the railway has made a deep cut. The top of this ridge commands a fine view up and down the Rio Grande, and there idle figures of Mexican or Indian are always to be seen watching for the train or studying the movements of almost invisible people on the other side of the valley. Draped in black, for the most part, motionless and immovable, they remind one irresistibly of Poe's picture:—

> "And the raven, never flitting, still is sitting, still is sitting,
> On the pallid bust of Pallas, just above my chamber door."

This point, as I have intimated, is in the midst of the civilization of northern New Mexico. Twenty-five miles up the Chama stands the large town of Abiquiu, an important place in old times; nearer by another plaza, Cuchillo, is a farming center. Not far away, in the Rio Grande valley, are San Juan, Santa Cruz and Española, the latter on the western bank of the river and the present terminus of the railway line southward, whence stages depart regularly for Santa Fe.

A Mexican farmhouse or "ranch" looks like a small fort, and makes a very pleasing picture, as you may observe in our sketch. It is square, rarely more than one story high, is built of mud, and roofed with immense round rafters, the ends of which protrude irregularly beyond the wall, because the builders have been too indolent to saw them off. Over these rafters,—above the line of which the wall extends a few inches,—are laid some boards or a stratum of poles, and upon these dry earth is spread a foot or more deep, with rude gutters arranged to carry away the water. In the course of two or three seasons, such a roof will have caught a supply of wind-sown seeds, and support a plentiful crop of grass and weeds, which is no disadvantage. This novel result is interfered with somewhat, however, by the habit of using the roofs of the houses (reached by a short ladder) as a place for drying fruit and sunning grain, and for a general lounging spot, whence a better view of what is occurring in the world,—the going and coming of the neighbors, the planting or gathering of the crops, the approach of a stranger-horseman, or the movements of the cattle on the benches,—can be obtained, than a seat on the ground affords. As the train dashes by, the passenger notices two or three women and children standing on each housetop, shading their eyes with their brown hands and making an unconscious *pose* irresistibly alluring to an artist.

On a line with the front of the house a wall will probably extend a little distance in each direction, and then backward, enclosing a garden and diminutive orchard. Everything is square. The idea of a curve seems rarely to enter the Spanish-Indian mind. For graphic effect, this is highly gratifying, since the bends in the river, the rounded outlines of the mountains, the undulations of foliage, are all in curves, to which the angular lines of the buildings present a most pleasing contrast. Now and then you will see a better house—one whitewashed outside, and having a balcony running around the second story. The outbuildings, in any case, are only a few mud huts, used for storage, and some rough pens where the animals are kept. Anything like the barns of an Eastern farmer is unknown.

The isolated dwelling, however, is largely a modern innovation. The general plan is to live in compact, block-like villages, surrounded by a wall, or what amounts to that. This results partly from the need in early days of united protection against the Indians, but chiefly from following the traditional custom of their red ancestors; for the New Mexican of to-day is a half-breed, or a mongrel of some degree between the Spaniard who "came over with the conqueror" and the Indian of whatever tribe happened to be accessible. Remote from civilized influences, the common people have tended always toward barbarous ways, and are more Indian than Spanish, albeit the dialect they speak is not so far removed from the Castilian as one would expect. There are local differences and idioms, of course, which are at once noticeable; but the usual tongue is not very bad Spanish.

Though Mexican hamlets and farms are scattered everywhere about here, in the fertile valleys, there is a class of towns along this part of the valley of the Rio Grande which are primarily Indian, and situated upon reservations each ten miles square, secured to them by the government. Each of these present villages, now commonly known by a Spanish name, was the site of an ancient native pueblo, and the fields which were deeded to them by the United States are those their fathers cultivated before the white men appeared at all. Some, however, yet retain their Indian names, as Taos, Picuris, Pecos, Pojuaque, Acoma and Tesúque. San Juan, Santa Cruz, Santa Clara, San Ildefonso, and others have been given new names by the conquerors and priests. South and west of Santa Fe lie many other pueblos, some of them very populous, as Jemez, Zia, Santo Domingo, and San Felipe. In all of them substantially the same sort of life is found, and as it is impossible for me to cover the wide territory they embrace, the reader must be content with the type of the whole to be seen here at Pueblo San Juan. It is a phase of humanity and conduct rapidly passing away—melting under the steady sun of modern progress; and the traveler who does not take an early opportunity to study it will miss not only that which is extremely interesting and suggestive, but what in a few years will become a matter of history and romantic tradition.

Here at Chamita the river is divided by a large, flat island into two branches, each perhaps a hundred yards in width. Over the first one, at the time of our visit, stood a good bridge, built by the railway company; a second bridge had spanned the other branch until the high water carried it away. Formerly there had been a ferry, but the boat

A PATRIARCH.

was out of order, and nobody cared to repair it, for could not the stream be forded?

In the cool of the evening the whole party went down to the river bank, trusting to good fortune for transportation. Thus challenged, good fortune stood by us in the person of a citizen and his broncho. Chartering the latter, the Artist, his sketching haversack slung over his shoulder, mounted, and then the Madame was invited to ascend, the pillion being a shawl thrown over the horse's haunches. When there she declared she could not stay—would certainly slip off the Gothic back of the beast the instant he moved.

"Then take it Anna Dickinson fashion," remarked her unfeeling spouse; whereupon there was a frantic lurch, a twinkle of crimson suspected to be hosiery, and a cheery "All right!" to let us know she was ready to brave the passage. The landing was safe, and then the patient horse returned and repeated the fording until we all were across.

But our peace of mind, or our *amour propre*—which is much the same thing—was disturbed by a suspicion that we were being laughed at, for a party of Mexican women from Chamita came down to the brink while we were there, and, chattering merrily over our slow and undoubtedly ludicrous progress, unconcernedly pulled off their shoes and stockings, gathered their skirts in a bunch about their waists, and gaily waded through as though in contempt of our fear of water and the conventionalities.

The large island was gravelly and liable to be inundated, so that it was given over to the pasturage of sheep, goats, cattle, horses and donkeys. On the eastern bank we found ourselves at once in the midst of grain fields and garden plats. Tall Indian corn alternated with short wheat, hay and alfalfa, or patches of potatoes, melons and vegetables. Fences were few, but the road was defined by a line of upright brush, bound into cohesion by withes of bark, so that it resembled a thoroughly dead hedge. Here and there stood a *casa chiquita*, but the main town was on the bluff marking the old bank of the river, half a mile from its present current.

Pueblo in Spanish simply means "a village." When the first explorers, Cabeça de Vaca, Coronado and the early lieutenants and friars whom Cortez sent northward, in search more of gold than geography, penetrated what is now New Mexico and Arizona, they everywhere found Indians more or less nomadic, but the larger part of the natives belonging to a different class, and living in settled communities of permanent houses. To these the Spaniards naturally gave the name of "village," or "pueblo" Indians, which, by a common process of lingual change, has become shortened into Pueblos, though Puebloans is a far better word. Their own tribal names have disappeared except in a few cases, such as the Zuñis and the Moquis, and the Spanish word covers all Village Indians distinguished from the roving Apaches, Mojaves and Utes, that surround them and centuries ago wrested from them much of their former territory. At present there are in all New Mexico but nineteen towns of the Village Indians, whose aggregate population in 1880 was only 10,469, as follows:

Taos	391
San Juan	408
Santa Clara	212
San Ildefonso	139
Picuris	1,115
Nambé	66
Pojuaque	26
Tesuque	99
Sochiti	271
San Domingo	1,123
San Felipe	613
Jemez	401
Silla (or Zia)	58
Santa Aña	489
Laguna	968
Isoleta	1,081
Sandia	345
Zuñi	2,082
Acoma	582

MAID AND MATRON.

Ascending the high bank along a road greatly gullied by the rains, we found ourselves in a large group of houses, each of which was joined to its neighbor as continuously as in a city block, but only one story high; or if there was a second story, it did not come out flush with the front wall, but was ten or fifteen feet back, the roof of the lower story serving as a portico to the upper floor, which was reached by an outside ladder.

These dwellings were built of mud bricks, called *adobes*, and in many cases the floors were lower than the level of the street—a matter of small concern, since the door-sill was so high as to shut out any water which might be running outside. Mixing in a little broken straw, rough blocks about twice the size of ordinary bricks are moulded, dried somewhat in the sun, and laid up in the form of a wall. Space is left for a door and some small holes for windows, quite high up. That is about all there seems to be of it, yet the inexpert find it not so easy to build a "doby" as they supposed. The consistency of the clay must be right, and I am told the wall must be laid so that the blocks somewhat brace each other by beveled sides, or else the great weight which rests on the top, otherwise wholly unsupported, will cause the middle of the wall to bulge. That these ancient houses stand so plumb and uncracked shows how proficient the Indians are at this peculiar architecture; and ought they not to be, for did not they invent it?

All the buildings are smoothly plastered outside and in. This is done some weeks after they are built, and after they have thoroughly dried. To obtain the necessary material for the outer "stucco" coat, the floor of the interior of the unfinished house is dug up and mixed with water until it becomes a soft paste. Then it is taken by the handful, dashed against the unchinked adobes, and spread smoothly with the palms, just as a town mason would use a trowel. The women do all this, and I remember surprising three damsels, as pretty as the New Mexican peasantry have to show, down on their knees and up to their elbows in seal-brown mud, plastering the new house, while father and mother were busy in the fields.

Most of the Indian dwellings,—and they are as good as the majority of the abodes of the Mexican ranchmen, — have two rooms, and sometimes three, but these are generally so dark that the eye must accustom itself to the gloom before their contents can well be discerned. This arises from the scarcity and diminutive size of the windows. Here in San Juan, indeed, I saw roughly sashed windows in many houses, or else a single pane of glass set in; but often only a grating is used to guard the aperture, or else holes in the walls are left so small that no enemy could crawl through. You can imagine the darkness inside, therefore, even on a bright day. Originally the pueblo was common property, and both men and women assisted in building it, but new ideas of individual possessions are invading the old notions. It was the former custom, too, to mix ashes with earth and charcoal into a substitute for mortar; yet, as we shall see later, the very ancient, ruined buildings of the ancestors of these Puebloans show an architecture in stone, with a cement now as hard, or even more tenacious, than the blocks it binds together. "They take great pride," says an old book "in their, to them, magnificent structures, averring that as fortresses they have ever proved impregnable. To wall out black barbarism was what the Pueblos wanted; under these conditions time was giving them civilization."

Entering one of the houses here in San Juan, we shall find the floor is only of earth, but that many skins are spread about. In one corner, or else beside the entrance door, will be one of the queer little round-topped fireplaces prevalent all over Spanish America; but if in the latter place, a low wall or wing of masonry runs out into the room, protecting the fire from contrary drafts. The cooking in summer is done out of doors almost wholly; but in cold weather, when utilizing these fireplaces, they use the iron pots and skillets which civilization has brought them, eking out with variously shaped earthen utensils of their own make, and baskets obtained from Apache and Navajo visitors.

You must expect to see very little furniture in an Indian's house, though occasionally some familiar objects are found. The beds are made on the floor, and consist entirely of skins and blankets. The walls are often whitewashed, and though they never heard of Eastlake, they always make a dado of clay water. The soft brown tint contrasts well with the white frieze, and would be attractive in itself; but the clay here is full of specks of mica, which dust the walls with gleaming points not to be spurned in mural decoration.

The Indians admire pictures, but are not scrupulous as to artistic superiority. In nearly every house you will find a board a few inches square, upon which is painted a religious subject, usually in red and yellow, of some saint, or a group of them. Such pictures, and others whenever they can get them, are highly valued and will be adorned with peacock feathers and bright berries.

They love gay colors and choose them in their dress, which is a singular mixture of Indian, Mexican and American. There go a man and woman ahead of us who are fair types. Neither are of large size, and though an oddity of gait comes from their habit of walking with their toes straight before them, both are of erect carriage. The man is dressed in brown flannel shirt, hanging blouse-like about him, tightly fitted leggings of buckskin, with a broad seam-flap in place of fringe on the outside of each leg, and moccasins. Over his right shoulder and under his left arm is loosely draped a striped blanket made by the Navajo or Apache Indians of the interior, and diligently repaired in its worn places. His head is bare, under the blaze of the hot sun, save for a wreath of cottonwood leaves. Under this "bay crown" his smoothly-brushed and jet black hair, accurately parted in the middle along a line of red or yellow ochre, is plaited on either side into two long braids, intertwined and lengthened out with strips of red flannel and tufts of otter-skin.

The woman wears a long, loose tunic of coarse cloth, almost devoid of sleeves, and belted at the waist; but sometimes this is of buckskin. Her extremities are not clad in leggings, but encased in short, shapeless boots having a moccasin foot, and stiff legs, which reach nearly to her knees, and often afford the only recognizable distinction between a male or a female, who, to a stranger's eye, are confusingly alike. She wears thrown over her head a shawl-like expanse of common pink-printed calico; but if you could see her hair you would discover

OLD CHURCH OF SAN JUAN.

that none of the attention had been bestowed upon it which her husband's has received; it has been cut short, particularly across the forehead, and is likely to be tangled and dirty. In this respect these Rio Grande Indians have fallen from grace into the slovenliness of their nomadic neighbors. The maidens of the purer Moqui pueblos, for example, take great care of their raven locks. Parting the hair at the back of the head, they roll it around hoops, when it is fastened in two high bunches, one on each side, a single feather being sometimes placed in the center. The Moqui wives gather it into two tight knots at the side, or one at the back of the head; and the men cut their hair in front of the ears and in a line with the eyebrows, while at the back it is plaited or gathered into a single bunch and tied with a band.

This woman is going to one of the public wells to draw water, and presently is joined by a young Hebe, with bare, shapely ankles and rotund bust, whose laughing talk is like the gurgling chatter of the blackbirds in the rushes. They each carry classically shaped and gaily ornamented jars of earthenware, made by themselves, and which they will tell you are *tinájas*. Some of the wells are so shallow that an inclined passage-way has been cut down to the water from the surface; from others the liquid must be drawn in buckets. Having filled their vessels, each woman lays a little pad on her head, skillfully poises the heavy tinája upon it, and marches off, as erect, elastic of tread and graceful in mien as any Ganymede who ever handed about the nectar on Olympus. You can see the trimmed and painted gourd-dipper floating about in the neck of the jar, and thus know that the water is level with the top; yet up hill and **down, along the dusty roadway,** through the half-concealing **corn, and** under the **low doorway go the** dusky carriers, and not a **drop is lost.**

A short distance **back we had met a superannuated governor, or chief,** called in Spanish Attencio. **His** long, straight hair, **of ashy hue,** and deeply furrowed features, gave a most venerable appearance to his attenuated but still upright form. His garments evinced more design, were better fitting, and **somewhat** fantastically decorated; while from his neck was suspended **a drum, a** tribute apparently to growing infirmities which had not quite obscured the dream of place **and circumstance.** We halted in curiosity while the Photographer, **by specious** argument and a gentle subsidizing process, overcame the half-scruples of the patriarch, and transferred his semblance to a "dry plate," an operation he repeated a little later with the maid and matron whom **we had seen at the well,** though in their case with more difficulty and overcoming of native shyness. The results of this enterprise are commended to the reader.

The pueblo pottery is of all sizes and shapes,—jars, pitchers, **canteens, bowls, platters, and** images of men **and** animals, made as **playthings for their children, or merely for** amusement, and the latter often **called their "gods" by ignorant** tourists.

It is evident everywhere that originally much finer **and more symmetrical pottery was made** by all these Village Indians than now. **They seem to have understood** the art of mixing a finer paste, and they worked with more careful hands. The resemblance of this **antique** ware to that of Egypt and Cyprus, has been noted in its structure, **and** in the "scrolls, straight lines and walls of Troy," with which it is embellished. Birds, too, were painted upon some of the oldest **ware extant,** recalling certain Chinese symbols, while "in the animal handles and in a design known as the old Japanese seal," the early ware of Japan is simulated. The ancient and (in ruins) most widely distributed form of pottery known is the "corrugated," fragments of which **are also** found in the mounds of the Ohio and Mississippi valleys and **on** the Pacific slope. This variety was made by winding **around** and above one another slender strings or ropes of red clay, expanding and con**tracting the coil to suit** the varying diameters of the vessel. Pressure **of the fingers alone, or aided** only **by a smooth stone, then** compressed **the coils into compactness and on** the inside into some smoothness. **There was also a kind** of ware in use in prehistoric times which bore a red or black glaze beyond anything seen in **later manufacture;** but this fine finish is thought to have been accidental.

At San Juan, as in all other pueblos, the old adobe church, with its absurdly barbaric furniture and uncouth appearance, is a center of interest. Climbing the rickety ladder to its little gallery, and thence ascending to the roof, one gets the best idea of how valuable a garden spot this district is. As far as the eye can reach, up and down the river, stretch farms and orchards and plazitas. I suppose that from the mouth of the cañon down to the village of San Ildefonso, a distance of about thirty miles, the river-bottom is almost continuously cultivated, the chief crops being wheat and Indian corn, the latter notable for its variegated and bright colors, and for which the people here keep the original name, *maiz*; but every sort of grain and vegetable is also produced in abundance.

The population sustained consists largely of Indians, in some localities, as here at San Juan, almost entirely so; and they are quite as industrious and skillful in their farming as the Mexicans. In most of the villages the tillage of the reservation is wholly in common, but here the Indians many years ago divided up their farming lands into individual properties, not all equally either, for it was apportioned to each man in proportion to his needs, abilities and desire. It is said that there has been little change in the ownership of this property, the same fields descending from father to son, generation after generation. This being the case, it is not strange to learn the second fact, that there is small variation in the fortunes of the different families, and that there is slight disposition on the part of any to become rich while others grow poor. All are self-supporting, and proud of the fact that no aid is asked or received from the government.

Nothing reminds the traveler more of the Holy Land than to witness these people threshing their grain, which happens, of course, in August, since they do not stack the grain in the straw at all. The threshing-floors are circular spaces of high level ground in the outskirts of the pueblo, around which poles ten or twenty feet high are set, though there is no need of more than mere posts. When the threshing is to be done, a rawhide rope or two is stretched about these posts to form a fence, and often upon this are hung many blankets, the gay colors and striped ornamentation of which make an exceedingly picturesque scene. Sometimes in place of the ropes a cordon of bare-legged small boys and girls, to whom the duty is great sport, does service as a girdle. The diameter of such a prepared space, hardened by service for half a century to the consistency of brick, is sixty or a hundred feet. In the middle of it are heaped the sheaves of the three or four families who are accustomed to join in this work until a suitable quantity has been obtained, and then the fun begins.

Through an opening in the extemporized fence is driven a flock of sheep and goats, or else a small herd of horses. They at once fall to eating the fresh grain, but are quickly beaten off and started into a run around the enclosure, trampling down the edges of the stack, and all the time getting more and more of it under their beating hoofs. Behind them race two or three athletic, bare-headed and scantily-dressed youths, cracking long whips, hustling the laggards, and nimbly keeping out of the way of the kicking, crowding and bewildered animals. This is quite as hard work as any of the horses or goats do, and is accompanied by continual halloos and trilling cries, which almost make a song when heard at a little distance. Now and then a young horse will make a leap at the rope, and snap the rawhide lariat, or dodge under it; or a venturesome goat will elude his guard and escape; but there are excited youngsters enough to speedily give chase and bring him back; and from time to time the panting drivers are changed, the animals given a rest, and the grain heaped into a new pile in the center. It is a wonderfully lively and gay picture, which will never be forgotten, and entirely unlike anything else to be seen in the United States. Toward evening, when the incessant tramping has threshed all the grains out of their husks, comes the winnowing. This is quite as primitive and idyllic as its forerunner. Having lifted away the bulk of the straw, several men and women take long-handled, flat-bladed wooden shovels, and toss up the grain which lies thick on the hard clay floor, thus allowing the wind to blow away the chaff. There is generally a breeze at sunset every day, and the largest part of the chaff is gotten rid of by the shoveling; but to perfect the process, the women take half a bushel at once of the grain, and re-winnow it, by tossing it a second time in and out of one of the large Navajo wicker-baskets, of which every family owns a number. The rough, wasteful threshing, and the cleansing, only partial at best, having thus been accomplished, the grain is divided out to its owners, and by them packed away in huge jars of coarse earthenware, called *ollas*, some of which will hold several bushels. These vessels keep it dry and safe from rats, so long as the covers are tight. All these processes are followed not only by the Indians, but by all Mexican farmers throughout the Spanish southwest.

We were never weary of wandering about these Indian towns, and watching the people at their work and sunny-tempered play. They are the happiest men and women on the continent. Well sheltered, well fed, well companioned, peaceful, guileless,—what else do they wish? Not theirs to know carking care, and the fluctuating markets which imperil hard-earned gains; nor to suffer the hurt unsatisfied ambition feels, or know the terrors of a crime-haunted or doubt-stricken conscience. The broad, bright sunshine of their latitude suffuses their whole lives and dispositions, turning their rock-bounded lowlands into a Vale of Tempe.

IX

SANTA FE AND THE SACRED VALLEY.

> Ages are made up
> Of such small parts as these, and men look back,
> Worn and bewilder'd, wond'ring how it is.
>
> —JOANNA BAILLIE.

I HAVE referred to Española as the southern terminus of the railway. From this point, however, another company is actively engaged in constructing a line to Santa Fe, a distance of thirty-four miles by the survey, and its prospective early completion will afford a direct and desirable connection with the ancient capital. At present the communication is by means of stages, which run in conjunction with the trains, and, not being restricted in the matter of grades, accomplish the trip in about twenty-five miles. The journey is interesting, and is made comfortably.

Santa Fe claims the distinction of being the oldest town in the United States, a claim that is readily admitted when we consider that it was a populous Indian pueblo when the first Spaniards crossed the territory now known as New Mexico, less than forty years after the discovery of the western continent by Columbus. The earliest European who penetrated this region was Alvar Nuñez Cabeça de Vaca, a Spanish navigator, whose vessel was wrecked on the coast of Texas in 1528, and who, with three of his crew, wandered for six years through the plains and mountains, until finally he joined his countrymen under Cortez in Mexico. His report of the section through which he passed led to an expedition, in 1539, by Marco de Niça, a Franciscan friar, who was frightened away by the Indians, and returned to Mexico with a marvelous account of the extent, population and wealth of the country, the magnificence of its cities, and the ferocity of its people. In 1540 the famous expedition of Vasquez de Coronado passed through the pueblo where Santa Fe now stands, crossed the range, and traversed the plains until he came to the Missouri river, at a point probably near the present site of Atchison or Leavenworth. In 1581, Friar Augustin Ruyz, with one companion, reached a village called Poala, a few miles north of Albuquerque, where they were killed by the Indians. Antonio de Espejo came with an expedition, in 1582, to seek Ruyz, and discovered Zuñi, Acoma and other pueblos. In 1595 Juan de Oñate founded a colony near the junction of the Rio Chama with

PUEBLO DE TAOS.

the Rio Grande, in the immediate vicinity of Española. It was about this date that a Spanish settlement was formed in Santa Fe, and the church of San Miguel erected. In 1680 there was a great uprising of the natives, who entirely drove out the Spaniards, and obliterated as far as possible all evidences of their occupation, dismantling, among other buildings, the old church. Twelve years later they were reconquered by Diego de Bargas. From that time to the present Santa Fe has had an eventful career. The Mexicans, in 1821, declared their independence of Spanish rule, and after that there were numerous insurrections, until the occupation of the territory by the United States, in 1846. Then came the War of the Rebellion, in 1861–65, in the course of which Santa Fe was captured by the confederates, and recaptured by the Union forces.

During all these years Santa Fe has changed its character but little, and is to-day, in general appearance, very much the same old Mexican town that it has been for nearly three hundred years. There is the same broad plaza, with the same adobe buildings nearly all the way around it; the same one-story houses, surrounding the same plazitas; the same suburban fields and gardens; and the same swarthy, dark-eyed population, still speaking the musical Spanish tongue. Wood is still brought into town on the backs of burros, and by this conveyance can be left inside the dwellings. Among the other objects of attraction to the stranger, are the governor's palace; the ruins of old Fort Marcy, on a bluff, from which is had a fine view of the city; and the extensive and beautiful garden of Bishop Lamy. The famous chapel of San Miguel, the oldest in America, still rears the same mud walls that have stood for three centuries, and internally is well preserved and in presentable shape. It has no exterior beauty and no interior magnificence, its only interest being in its age and the sacred uses for which it has been kept up during almost the entire period of American civilization. On a great beam, as plain as if made but yesterday, is the Spanish inscription, traced there one hundred and seventy-four years ago, to the effect that "The Marquis de la Penuela erected this building, the Royal Ensign Don Augustin Flores Vergara, his servant, A.D. 1710." Original documents show that this refers to its restoration after the wood-work was burned by the rebel Indians. A dark picture of the Annunciation, on one side of the altar, bears on its back a notation, seemingly dated A.D. 1287, leading to the belief that it is one of the oldest oil paintings in the world. By the side of the church is a two-story adobe house, which tradition says was in existence when Coronado marched through the town. The neighborhood of Santa Fe is rich in precious stones, including turquoise, bloodstone, onyx, agate, garnet and opal. The manufacture of Mexican filigree jewelry, largely carried on here, will be found interesting. The work is done by natives, to whom the trade has been handed down by their ancestors, who derived it from the Italians. The primitive Spanish records of the aborigines of all tropical America say that there were "no better goldsmiths in the world;" so that the Indian blood mixed in the veins of most of the modern artisans may have increased their skill.

But even quaint old Santa Fe is catching the spirit of the age, and now boasts a colony of northern residents, cultivated society, and many handsome structures, among which are a new hotel, a large public hospital built of stone and brick, a Methodist church, Santa Fe Academy, and San Miguel College. The tendency of this innovation will be to rapidly dissipate the aroma of antiquity and sentiment which has hitherto attached to the town's romantic history.

On the return northward, our cars were again set out at Embudo, the gentlemen of the party having determined not to omit from this itinerary the Taos pueblos, possibly the most antiquated, and certainly the best preserved of all, and whose people are still awaiting with pathetic patience the returning Montezuma, who shall restore their pristine glory, and the kingdom that stretched from river to sea. When questioned as to her desires, the Madame did not advance the staple feminine excuse, a council with the dressmaker, but boldly proclaimed her aversion to the thirty-mile equestrian trip. So we left her behind reluctantly, with many injunctions to our *chef de cuisine* and still more trustworthy railway friends.

After no little wrangling, a sufficient number of spiritless quadrupeds were procured from the natives, and we turned our faces to the north. And right here let me advise the reader who may hereafter contemplate this pilgrimage, to address, in ample season, Mr. Henry Dibble, Fernandez de Taos, New Mexico, who will undertake to have a team in waiting at Embudo, thus saving the traveler much time and even more bad Spanish. The ride was thoroughly enjoyable, and formed a fitting prelude to the novel experiences that followed. I have said "ride," though the statement is not altogether exact, since we took pity,—and largely from necessity,—on our miserable under-sized and under-fed ponies and ourselves, and walked a full third of the distance. Occasionally we passed small Mexican villages, which seemed as peacefully asleep in the afternoon sunshine as we could ever have pictured them. Flocks of ugly yellow-spotted goats, attended by dusky urchins in scanty attire, browsed on the near hill-slopes. Over the eaves of almost every one of the low adobe houses hung great ropes of red peppers,—the *chili colorado* of the Mexican,—that gave the one brilliant dash of color to a perspective whose tones were otherwise the most subdued. Not unfrequently, however, an entire family would be seen ranged along in a row on the shady side of the house, the women generally dressed in gay colors, solid red, blue, or green, and all as silent as the scene on which they looked. But for the dogs, these hamlets might have passed for the ruins they appeared. The *caretta*, or great, clumsy, two-wheeled cart, and the plow made of a pointed stick, were here and there reposing before the abodes, not seeming like implements of daily use, but as appropriate details in the worn-out landscape.

To pick one's way through Taos valley, even by daylight, might be a task; in the darkness, which at length overtook us, success was a lucky chance. The very populousness of the locality was against us. How many times we took the divergent road and brought up against the fence of some ranchero's threshing-floor; or crossed the stream *not* at the ford; or engaged in a broil with some awakened native who persisted in misunderstanding our gesticulated inquiries, may never be related. The houses, too, were a mystery. To find the front door of the rectangular heap of mud; to determine in what part of its cavernous recesses the inmates might now be residing; or to decide whether, after all, it was not the stable, taxed our ingenuity and tempers through several hours of that memorable evening. Finally, in the plaza of a great communistic ranch-house, that covered an area half as large as a city block, we managed to secure the services of a *muchacho*, who preceded us on horseback, and led us into one of the narrow and crooked streets of Fernandez de Taos, where we soon found the hostelry of "Pap" Dibble.

Taos valley is widest near its head, where the several streams that form the river issue from the Culebra range. About the center of the fertile expanse lies the old Mexican town of Fernandez de Taos, with a present population of 1,500. Two miles northeast of it, and under the shadow of Taos mountain, stand the two great buildings known as the Pueblo de Taos, and inhabited by about 400 Indians. Three miles south of Fernandez lies still another Mexican village, named Ranchos de Taos, in contrast with whose adobes the traveler finds a newly-erected flouring mill. The middle settlement has the greatest commercial importance, and is likewise possessed of considerable historic interest. Here was the seat of the first civil government of the territory by the United States, after it had been acquired as a result of the Mexican war of 1846. Here Bent, the first governor, was killed in the revolt of the following year, and the ruins yet remain of the old church on whose solid mud walls the howitzers of the troops could make no impression, and from which the band of insurgent Indians and Mexicans were only finally dislodged by means of hand-grenades. The widow of the murdered governor still lives in her modest adobe, and shows to visitors the hole in the wall made by the fatal bullet. Fernandez de Taos was likewise for years the residence of Colonel Kit Carson, and in the walled graveyard at the edge of town his body is buried.

All this and much more is communicated the following morning by the genial Dibble, who fills our idea of what a host should be. For twenty years has he lived in this quiet valley, among an alien people, leaving it only once for a trip to Santa Fe, content to preside over his curious aggregation of rambling adobes, and make each chance guest feel himself under paternal care.

But to us the great interest centers in the Indian carnival at the pueblo, which is to occur on the morrow. On the last day of September of each year the Taos Indians celebrate the festival of their patron saint, San Geronimo (the Spanish St. Jerome), by ceremonies altogether unique, and which few Americans have as yet witnessed. Some hours are still at our command, in which to study the country in its every-day aspect, and we early start out in the direction of the pueblo. Already the roads converging toward the old stronghold show signs of the assembling throng. Little bands of Indians, gaily blanketed and with uncovered heads, who have walked from pueblos perhaps fifty miles away, driving before them shaggy burros, with many-shaped packs; and Mexican fruit-vendors, their trains of donkeys laden with well-filled wicker baskets, form the vanguard of the unique procession. The valley across which we pass is all under cultivation, and the ground is now covered with the yellow stubble, while along the roadside we come upon the regulation threshing-places.

The Pueblo de Taos consists of two great mud buildings (of the larger of which an engraving is given) facing each other from opposite banks of a stream, and perhaps two hundred yards apart. They rise to a height of about fifty feet, and seem to have attained their present size by accretions during the ages since they were founded. They are of an irregular pyramidal form, and made up of about five stories or terraces. Each new story is built a distance from the edge of the one immediately beneath, so that both the length and breadth of the building diminish as the height is increased. To enter the rooms we must ascend one of the many ladders that lean against the wall, and then descend another ladder through a hole in the roof. Everything was quiet and silent about this great human wasps' nest. Nude children tumbled on the ground in the warm rays of the sun; men strolled lazily hither and thither, their bodies wrapped in gaudy blankets and legs encased in close-fitting sheepskin leggings, while to their hair, black as jet and brought down in a lock on each side, hung great bunches of zephyr or other gay material; women, dressed in much the same manner, carried on their heads the earthen water-jars, or large baskets of bread, which had been baked in the oval mud ovens ranged in front of the pueblo. Everybody treated us with quiet respect, and seemed pleased to respond to our salutations. We climbed over one of the ancient piles, mounting to its topmost story on shaky ladders, peering into its rooms, which we were courteously invited to enter, and where we found sometimes as many as a dozen Indians sitting on the floor, engaged in adding some last touches to the holiday garments. We saw few young men, but afterwards learned that they were in the *estufas*, or underground council-chambers, preparing for the next day's spectacle.

To give anything like an adequate account of the festival would require a small volume. Early on that resplendent September morning the human tide began to pour in, till, from our position on the summit of the north pueblo, we looked down to the plaza below on a surging mass of fully three thousand Indians and Mexicans, in every gay and fantastic garb. The fruit-vendors had established themselves in scores of little stalls scattered over the plaza, and with their burros standing patiently by, added a picturesque feature to the scene. Three hundred mad young Mexicans, mounted on excited ponies, charged among the crowd in a body, dared each other in feats of horsemanship, or "ran the *gallo*" on the opposite bank. The padre from Santa Fe first held service in the little church, after which came the event of the day. One hundred naked and painted Indians issued in solemn march from an *estufa*, and began the race, two by two, over the straight track a thousand feet long. For an hour and a half they sped up and down in front of the pueblo, amid the wildest excitement of the spectators. Then the march of the victors, to the music of a wild chant, while bread is showered upon them from one of the roofs of the pueblo under which they pass, closes the morning's ceremonies.

The afternoon is consumed by the antics of seven unclothed and curiously painted clowns. For three hours do they amuse that motley crowd

PHANTOM CURVE.

with their mimic cock and bull fights, and their semblance of plowing, threshing, and other familiar labors. As the sun nears the west, the rabble gather about a pole, fifty feet high, over the cross-piece at whose top has been hung a living sheep, together with garlands of fruit and a basket of bread. After many pretended failures, the pole is climbed, and the bread and fruit are thrown to the ground. Last of all, the sheep, in which a spark of life still lingers, is detached, and strikes the earth with a sickening thud. With yells and strange cries the Indians rush in, the sheep is torn limb from limb, and with this, the only revolting part of the entire celebration, the *fête* ends.

The lava-caps away down the valley were glowing golden as we rode back to Fernandez. Thought was busy with the strange events of the day. During how many centuries had these onlooking hills witnessed the gathering throngs of such festivals, since were laid the foundations of those dusky piles, now bathed in sunset glory, where tradition says the cultured hero Montezuma was born, and whence he set out on his prophetic career? And can this ancient people long withstand the civilization that is fast bearing down on them; or will it not soon engulf them and fill with modern life the sacred valley?

On our arrival at Embudo we found the Madame in much tribulation. Not that any harm had befallen her; but the cook, from being an assistant after a fashion, had immediately on our departure developed into an absolute dependent. This personage had for some time been a subject of much solicitude and serious discussion in our family circle. We could sympathize with his infirmities, but when they became the ever-present shield to the most aggravated laziness, our philosophy weakened. And so, when the Madame had explained his apparently total collapse, our decision was speedily reached. In spite of his protestations and his phenomenal physical improvement, we lifted him by main force on to the first train, and shipped him northward without our blessing.

Concerning this Amos, the Madame wrote as follows to her friend, Mrs. McAngle: "He was the 'Knight of the Sorrowful Countenance,' in our vocabulary of nicknames. His face was a suggestion of martyrdom done in coffee color, for he was a darkey of uncertain and speckled hue; and his religious possessions, a couple of books of devotion of the most melancholy kind, kept him up to his model. Everything had been put into our kitchen-car before leaving Denver, pell-mell; and when, at our first evening's halt, I went out to investigate, I found Amos sitting on a soap-box, in the midst of a chaos of utensils and packages of provisions, almost weeping at the water-splashed confusion, without making the least movement toward straightening matters. He brought with him two encumbrances,—a fifteen years' experience on the Sound steamer *Bristol* (so he said), and his Rheumatism, with a very big R.

"He was a good enough cook when he tried to be, but wholly averse to neatness. Becoming tired of seeing things that bore no relation to one another on an intimate acquaintance, as the bacon and flour, for instance, I undertook, with fear and trembling, some mild expostulation. But I had not gone far before he raised himself to all his dignity, and exclaimed, 'I have served fifteen years as first cook on board the *Bristol,*' and then turned his back upon me. Somewhat stunned, but persevering, I continued meekly to tell him the things I wished him to attend to. Instantly his tone changed from indignation to supplication, and he described in feeling terms his rheumatism. 'He enjoyed a neat kitchen as well as anybody, but what could he do, having his joints all knotted up with this terrible disease?' and his face grew sadder than ever. I retired from the field vanquished, and reported progress.

"The gentlemen were not so easily silenced, however, and that very day began a little investigation. 'Amos, can you make a tapioca pudding?' cried one, at lunch. 'I have been fifteen years chief cook on the *Bristol*,' came the answer, with an upward roll of the prayerful eyes. A little later: 'Amos, bring up a pail of fresh water from the creek.' Very glad to oblige you, sir (a groan), but I've the rheumatics.' When one excuse wouldn't answer the other would. So we sent him off, and got Burt in his place,—a youth without rheumatism or record,— who proved to be a very bright, willing, and useful boy."

X

TOLTEC GORGE.

> I'll look no more;
> Lest my brain turn, and the deficient sight
> Topple down headlong.
> —KING LEAR.

AVING at last turned our heels reluctantly on the simple-hearted, prettily-chequered life of the Pueblos, we raced back in a single night to the plains of San Luis. A long line of telegraph poles stretches out from Antonito into a true vanishing point across the park, and the train follows it San Juanward. The noble Sangre de Cristo looms up higher and higher behind us as we proceed, a mirage lifting the line of cottonwoods along the Rio Grande into impossibly tall and spindling caricatures of trees; while the Jemez mountains away to the south are not yet lost to view, and the striking landmark of Mount San Antonio, smooth and round, is close at hand. A few miles beyond it the arid level of the lake-spread plain breaks into white, stony eminences, reared in a bold front. To surmount these the track is arranged in long, ingenious loops, in one place, known as the "Whiplash," extending into three parallel lines, scarcely a stone's throw apart, but disposed terrace-like on the hillside. On top of the mesa the sage-brush disappears, grass, piñons and yellow pines taking its place, and we begin to wind among the long, straight lava ridges at the foot of the divide between the Los Pinos and the Chama, whence the backward view is remarkably fine. The road here is like a goat's path in its vagaries, and wagers are made as to the point of the compass to be aimed at five minutes in advance, or whether the track on the opposite side of the *crevasse* is the one we have just come over, or are now about to pursue.

Describing a number of large curves around constantly deepening depressions, we reached the breast of a mountain, whence we obtained our first glimpse into Los Pinos valley; and it came like a sudden revelation of beauty and grandeur. The approach had been picturesque and gentle in character. Now we found our train clinging to a narrow pathway carved out far up the mountain's side, while great masses of a volcanic conglomerate towered overhead, and the face of the opposing heights broke off into bristling crags. The river sank deeper and deeper into the narrowing vale, and the space beneath us to its banks was excitingly precipitous. We crowded upon the platform, the outer step of which sometimes hung over an abyss that made us shudder, till some friendly bank placed itself between us and the almost unbroken descent. But we learned to enjoy the imminent edge, along which the train crept so cautiously, and begrudged every instant that the landscape was shut out by intervening objects.

To say that the vision here is grand, awe-inspiring, painfully impressive or memorable, falls short of the truth in each case. It is too much to take in at once, and we were glad to pause again for a little brain-rest at a telegraph station, hung almost like a bird's nest among the rocks,— to grow used by degrees to the stupendous picture spread before us. We were so high that not only the bottom of the valley, where the silver ribbon of the Los Pinos trailed in and out among the trees, and underneath the headlands, but even the wooded tops of the further rounded hills were below us, and we could count the dim, distant peaks in New Mexico.

Six miles ahead lay the cañon of which we had heard so much,— the Toltec Gorge, whose praises could not be overdrawn. Evidently his majesty had entrenched himself in glories beside which any ordinary monarch would lose his magnificence. Was this king of cañons really so great he could afford to risk all rivalry? Here, on the left, what noble martello-tower of native lava is that which stands undizzied on the very brink of the precipice? I should like to roll it off and watch it cut a swath through that puny forest down there, and dam up the whole stream with its huge breadth. How these passages of spongy rock resound as our engine drags the long train we have again mounted through their lofty portals! How narrow apparently are these curved and smooth embankments that carry us across the ravines, and how spidery look the firmly-braced bridges that span the torrents! All the way the road-bed is heaped up or dug out artificially. It is merely a shelf near the summit. It hugs the wall like a chamois-stalker, creeping stealthily out to the end of and around each projecting spur; it explores every in-bending gulch, boldly strides across the water-channels, and walks undismayed upon the utmost verge, where rough cliffs overhang it, and the gulf sinks away hundreds of feet beneath.

In the most secluded nook of the mountains we come upon Phantom Curve, with its company of isolated rocks, made of stuff so hard as to have stood upright, tall, grotesque, and sunburned, beside the pigmy firs and cowering boulders with which they are surrounded. Miles away you can trace these black pinnacles, like sentinels, mid-way up the slopes; but here at hand they fill the eye, and in their fantastic resemblance to human shapes and things we know in miniature, seem to us crumbled images of the days when there were giants, and men of Titanic mold set up mementoes of their brawny heroes,—

"Achaian statues in a world so rich!"

Phantoms, they are called, and the statuesque shadows they cast, moving mysteriously along the white bluffs, as the sun declines, are uncanny and ghost-like, perhaps; but the brown, rough, grandly grouping monoliths of lava themselves, are no more phantoms than are the pyramids of Sahara, and beside them the Theban monuments of the mighty Rameses would sink into insignificance.

Winding along the slender track, among these solemn forms, we approach the gorge, the vastly seamed and wrinkled face of whose opposite wall confronts us under the frown of an intense shade,—unused to the light from all eternity; but on this, the sunny side, a rosy pile, lifts its massive head proudly far above us, its square, fearless forehead,—

"Fronting heaven's splendor,
Strong and full and clear."

How should we pass it? On the right stood the solid palisade of the sierra, rising unbroken to the ultimate heights; on the left the gulf, its sides more and more nearly vertical, more and more terrible in their armature of splintered ledges and pike-pointed tree-tops,— more often breaking away into perpendicular cliffs, whence we could hurl a pebble, or ourselves, into the mad torrent easily seen but too far below to be heard; and as we draw nearer, the rosy crags rise higher and more distinct across our path. We turn a curve in the track, the cars leaning toward the inside, as if they, too, retreated from the look down into that "vasty deep," and lo! a gateway tunneled through,—the barrier is conquered!

The blank of the tunnel gives one time to think. Pictures of the beetling, ebony-pillared cliffs linger in the retina suddenly deprived of the reality, and reproduce the seamed and jagged rocks in fiery similitude upon the darkness. In a twinkling the impression fades, and at the same instant you catch a gleam of advancing light, and dash out into the sunshine,— into the sunshine only? Oh, no, out into the air,— an awful leap abroad into invisibly bounded space; and you catch your breath, startled beyond self-control!

Then it is all over, and you are still on your feet, listening to the familiar ring of the brown walls as they fly past.

What was it you saw that made your breathing cease, and the blood chill in your heart with swift terror? It is hard to remember; but there remains a feeling of an instant's suspension over an irregular chasm that seemed cut to the very center of the earth, and, to your dilated eye, gleamed brightly at the bottom, as though it penetrated even the realms of Pluto. You knew it opened outwardly into the gorge, for there in front stood the mighty wall, bracing the mountain far overhead, and below flashed the foaming river. This is the sum of your recollection, photographed upon your brain by a mental process more instantaneous than any application of art, and never to be erased. Gradually you con-

PHANTOM ROCKS.

clude that the train ran directly out upon a short trestle, one end of which rests in the mouth of the tunnel, and the other in the jaws of a rock cutting. This is the fact; but the traveler reasons it out, for he cannot see the support beneath his car, which, to all intents, takes a flying bound across a cleft in the granite eleven hundred measured feet in depth.

Our train having halted, the Artist sought a favorable position for obtaining the sketch of Toltec Gorge which adorns these pages, the Photographer became similarly absorbed, and the remaining members of the expedition zealously examined a spot whose counterpart in rugged and inspiring sublimity probably does not exist elsewhere in America. A few rods up the cañon a thin and ragged pinnacle rises abruptly from the very bottom to a level with the railway track. This point has been christened Eva Cliff, and when we had gained its crest by dint of much laborious and hazardous climbing over a narrow gangway of rocks, by which it is barely connected with the neighboring bank, our exertions were well repaid by the splendid view of the gorge it afforded.

Just west of the tunnel, and close beside the track, the rocks have been broken and leveled into a small smooth space, and here, on the 26th of September, 1881, that gloomiest day in the decade for our people, were celebrated as impressive memorial services for GARFIELD, the noble man and beloved president, then lying dead on his stately catafalque in Cleveland, as were anywhere seen. The weather itself, in these remote and lonely mountains, seemed in unison with the sadness of the nation, for heavy black clouds swept overhead, and the wind made solemn moanings in the shaken trees. It was under circumstances so fittingly mournful that an excursion party, gathered from nearly every state in the Union, paused to express the universal sorrow, and to conceive the foundation of the massive monument which catches the traveler's eye on the brink of the gorge, and upon whose polished tablet are engraved these words :

IN MEMORIAM.

JAMES ABRAM GARFIELD,

PRESIDENT OF THE UNITED STATES,
DIED SEPTEMBER 19, 1881,
MOURNED BY ALL THE PEOPLE.

Erected by Members of the National Association of General Passenger and Ticket Agents, who hold Memorial Burial Services on this spot, September 26, 1881.

XI

ALONG THE SOUTHERN BORDER.

*There in the gorges that widen, descending
From cloud and from cold into summer eternal,
Gather the threads of the ice-gendered fountains,—
Gather to riotous torrents of crystal,
And giving each shelvy recess where they dally
The blooms of the north and its evergreen turfage.*

—BAYARD TAYLOR.

THOUGH the climax of the pass to the sight-seer is Toltec Gorge, the actual crest of the Pinos-Chama divide is at Cumbres, some fifteen miles westward, and several hundred feet higher. After leaving Toltec, the brink of the cliff is skirted for some time, and many grand and exciting views are presented; but the stream is broken into cascades, and rapidly rises to the plane of the track. Passing a number of snow-sheds, the train is soon twisting around shallow side ravines, and at last, after making a great circle of nearly a mile, there comes a stoppage of that dragging sensation which the wheels impart on an upward grade, and the cars halt on the little level space at the summit. From Antonito to Cumbres the maximum ascent to the mile is only seventy-five feet, while on the western slope the descent per mile reaches two hundred and eleven feet. This intrepid railway crosses the main ranges of the Rocky Mountains over seven or eight distinct passes; and in every instance the locating engineers have followed one water-course upward to its head, and another downward to the valley, finding invariably the sources of these oppositely-flowing brooks to be in springs only a few feet or rods apart at the top. In the present case so slight is the separation that we seem to stop beside the Los Pinos, and to start beside Wolf Creek. Although at an altitude of about 9,500 feet, the neat station buildings at Cumbres are located in a depressed indentation, whence the surrounding hills shut off all outlook.

Our train is scarcely in motion again, however, ere a deep gully opens at our feet, and we commence to crawl cautiously around the protruding face of Cumbres Mountain, with its curiously-piled top of red and gray sandstone, and its precipitous front, in which is hewn midway a shelf for the track. Beyond this we pass a great curve, and then overlook a beautiful valley, which leads down into the broad basin through which the Rio Chama pursues its way southeasterly to its junction with the Rio Grande at Chamita. The view here is picturesque, and well worthy the reproduction our artist has seen fit to give it. There are glimpses of far-off, white-edged mesa-lands, with spaces of shadowy cobalt between. The brook sinks deeper, and its grassy banks are full of yellow and purple asters, in brightest bloom, glorifying the whole hillside up to where, a short distance from its bed, begins the solid spruce and aspen forest. Near Lobato, the track crosses from one tawny ridge to another, on a lofty iron bridge, and we note that Wolf Creek is here a lovely stream, with many cozy nooks in which the sportsman may pitch his tent, and are informed that the water is full of trout, while the wooded mountain slopes abound in large and small game. Once down in the valley, the way is through smooth lawns and pleasant groves until Chama is reached, and here we pause to ask questions about sheep.

Our cars were set aside in the very woods, far from the noisy station; a Y runs southward there, the germ perhaps of a railway down the river to Chamita, where it may join the southern line. All about us are the never-silent pines, and the breezes that whisper among their rugged branches blow laden with balsamic odors. Close by is the Rio Chama, hidden between dense and continuous thickets, through which the cattle can tell you of winding and mysterious paths. Everything in the landscape is soft and peaceful. The grass lies green and tender; the rounded clusters of willows, blending with the glowing masses of poplar behind them, bright in their new autumn colors, make no sharp line against the pine copse, nor this against the swelling, gaily-clothed background of the hills above.

Through this utterly wild, yet richly modulated scene, the Madame and I rode off one morning down to Tierra Amarilla, leaving our companions to angle for finny beauties. For miles the two mules trotted gaily with us through alternate groups of gigantic yellow pines and open stretches of grassy upland, where now and then we struck panic into the hearts of a flock of sheep. Then signs of ranch-life began, and some cattle were met; and ten miles from Chama we came upon the thrifty plazita of Los Brazos (the Arms), surrounded by a wide district of farming land. This continued three miles, and centered in a second hamlet, Los Ojos (the Springs), where there were several shops; thence two miles more, across a sage-brush terrace, took us to our destination.

Though the post-office has restricted the use of the name to this village, the whole region, on account of its peculiar beds of ochre earth, was formerly known as La Tierra Amarilla. This has been abbreviated, not only in spelling, but in speaking, until its ordinary pronunciation is Terr-amaréea. In 1837 a tract forty miles square, in this part of the valley of the Chama, was granted by the Mexican government to Señor Manuel Martin and his eight sons. There was a failure to ratify the matter somehow, and in 1860, old Manuel having died, his eldest son, Francisco Martin, applied to the Surveyor-General of the United States to have the grant confirmed to him, his brothers "and their companions" resident thereon. The Surveyor-General, however, struck the "companions" out, and ratified the grant only to the heirs of Manuel Martin. When this was discovered, Francisco, with the consent of his brothers, at once gave to each incumbent the land he occupied, and a deed for the same. Soon after this the Martins sold out all the domain, getting scarcely ten thousand dollars for the whole million of acres, which passed chiefly into the hands of a gentleman of Santa Fe. The next proprietor is to be an English company, which proposes to colonize the tract with British farmers and stock-raisers. The price paid, it is said, amounts to more than two millions of dollars. Pending these successive arrangements, the unfortunate settlers found their deeds valueless, because of informality,—a neglect not at all strange in a Mexican hidalgo. On the point of being ousted of their supposed proprietary rights, if not actually dispossessed, they bethought themselves of a lucky law of the territory, which gives ownership to anyone who can show a color of title, and undisputed possession for ten years. This statute saved them, and they will be bought out by the Englishmen.

Farming here hardly yields enough of grain to meet the local demand, except in the oat crop. The soil is good, the irrigating facilities very large and convenient, timber is plenty, and the climate superb. Yet only a portion of the wide, fertile bottom-land is under cultivation, and the valley invites intelligent immigration with an array of inducements unusual in New Mexico.

But there is no laxity in the matter of wool-producing, a full million of sheep belonging at Tierra Amarilla, distributed among about two hundred owners. These are never sold, except under stress of need for money, when they bring from one to two dollars each. The value of the total flock, then, will be somewhat over a million of dollars; while the annual production of wool will amount to more than two millions of pounds, worth more or less than half a million dollars, according to the price of wool. Its natural outlet to market is through Chama and Amargo. Early in September the flocks are started on their march to the southern part of the territory, where they can feed unharmed by winter storms

I do not know a better place to study the primitive life of New Mexico, with all its quaint features; and the traveler who follows our example, and digresses long enough to ride down to the settlements I have mentioned, will not regret his short divergence from the beaten track.

Resuming the iron trail westward from Chama, all the way to Willow Creek the same beautiful parks of yellow pine continued, and the track crossed and recrossed a sparkling brook. Passing the mines of excellent bituminous coal at Monero, and surmounting a low water shed, which is in reality the continental divide, the deeply-notched tops of the

Sierra Madre came into view in the north, and we spanned the first of the many streams that flow down from it into the Rio San Juan. A birds eye view of this well-wooded and almost flat region, just on the line between Colorado and New Mexico, would have shown it to consist of a series of low, slightly-tilted ridges, parallel with which ran the serpentine and deeply-sunken rivers.

The first or easternmost of these streams is the Rio Navajo, encountered near Amargo, and up to which, all the way from Chama, nothing is to be found save grazing land, devoted mainly to sheep. Though its bottoms available for agriculture are probably broader than the water it contains is able to irrigate, far more farming remains to be done here than has yet been undertaken. The Rio San Juan, into which the Rio Navajo empties just west of Juanita, is the great drainage channel of this portion of Colorado and New Mexico, and a river of power even here. Its crystal-clear waters to-day prattle innocently, but they sometimes come down from the heights like an Indian raid, a besom of destruction for anything not as firmly anchored as the granite buttresses of the hills themselves.

From Amargo,—there is no end of bloody history attached to El Amargo and its fine cañon, dating from the early days of settlement, Indian fighting and border ruffianism,—runs the old stage-road northward to Pagosa Springs, Animas City, and the interior mines. The tales of that thoroughfare would furnish a whole library of flash literature without going much astray from the truth.

Pagosa is the far-famed "big medicine" of the Utes,—the greatest thermal fountains on the continent. "The largest of these springs is at least forty feet in diameter, and hot enough to cook an egg in a few minutes. Carbonic acid gas and steam bubble up in great quantities from the bottom, and keep the surface always in a state of agitation. The water has the faculty of dividing the light into its component colors, producing effects very similar to those of the opalescent glass of commerce. Around the large spring, and extending for a mile down the creek, are innumerable smaller ones, many of which discharge vast amounts of almost boiling water. These, being highly charged with saline matter, have produced by deposition all, or nearly all, of the ground in their vicinity, and their streams meander through its cavernous structure, often disappearing and reappearing many times before they finally emerge into the river. This spot must become a great popular resort. Its plentifully timbered and mountainous surroundings enhance the interest it otherwise possesses for the traveler and health-seeker, and the medicinal value of the springs claims the attention of all who can afford time to visit them.

"The village of Pagosa Springs is situated about four miles south of the base of the San Juan range, upon the immediate southeastern bank of the Rio San Juan. It consists of a group of dwellings, stores, and bath-houses, among which the steam of the hot springs issues in such clouds as at times to render the entire place invisible. Immediately above the town, on the opposite side of the river, rises a flat-topped, isolated hill, whose summit contains a plateau large enough to liberally accommodate the government post which has been erected there. Utilizing the pines so abundant in the neighborhood, the buildings are all made of logs; and model log-houses they are. A more inviting military camp, both as regards location and construction, could not well be conceived."

Pagosa lies in the heart of that splendid pine forest, which covers a tract one hundred and thirty miles east and west by from twenty to forty miles north and south. Here the trees grow tall and straight, and of enormous size. No underbrush hides their bright, clean shafts, and, curiously enough, it is only in special locations that any low ones are to be found. These monarchs of the forest seem to be the last of their race, and, like the Indians, are doomed very soon to disappear. They are of immense value, for they form a huge storehouse of the finest lumber in a country poorly supplied in general with such material.

The vicinity of the springs is destined to yield large crops under irrigation, though at present there is little settlement there. Mexicans pasture their sheep as thickly as the fields will hold them; and try to give their flocks a few days in the basin at least once each season, believing that the drinking of the waters is of great benefit to the animals. Though the upper valley of the San Juan is unlikely to prove very profitable as agricultural land, the lower parts, in New Mexico, are the scene of extensive and highly successful Indian farming operations. The next stream westward, however, the Rio de las Nutrias (River of Rabbits), has good ranches, and so has the Rio de las Piedras (Stony river), the Rio Florida (River of Flowers), the Rio de los Pinos (Pine river), and the Rio de las Animas Perdidas (River of Lost Souls), up whose valley we turned sharply when a few miles from Durango. But thus far only a fraction of the tillable soil has been located on.

At Amargo,—for in this sketch of the rivers I have run ahead of our actual progress,—we find several hundred Apaches waiting to receive their rations, it being the weekly issuing day. Three of the redskins importune us for a ride, and we take them upon our platform, having entomological objections against offering them the hospitalities of the interior of the car. Our fund of Spanish is mutually limited, but one of us has a fair knowledge of the sign language, learned in former wanderings among the Dakotas and Kalispelm; and while these Apaches never heard of either of those great northern nations of red men, they readily understand most of the signs, though frequently showing us with great good nature that their way of expressing an idea is by a somewhat different gesture.

Our visitors were men of medium size, beardless, and very dark.

TOLTEC GORGE.

Their hair was coal black, straight, parted in the middle, carefully combed, and gathered into two braids, the end of each being ornamented with a feather or a tuft of yarn. They wore woolen shirts, the original colors of which were lost in dirt; buckskin leggings, with fringes on the outer seam; moccasins of poorly tanned sheepskin, pointed at the toe and decorated with fringes. Bright scarlet blankets, marked U. S. I. D., were wrapped around their waists or drawn over their hatless polls. Each man carried a sheath-knife at his belt, and a bow with about a dozen arrows wrapped in a sheepskin case. Their features expressed much intelligence and good humor, easily breaking into chuckles of laughter, for they enjoyed studying us quite as much as we did them.

These Indians were Jicarilla Apaches, another branch of what was originally the same great tribe being the Mescalero Apaches, of southern New Mexico. The Jicarillas number about eight hundred souls, all told, and are apportioned into five bands, under as many chiefs, the most influential of whom is *Huarito* (Little Blonde), though he has no nominal headship. Their reservation extends thirty-three miles southward from the Colorado line, and is sixteen miles in breadth. On account of the severity of the winters about Amargo, the Government moved these Indians, during the autumn of 1883, to Fort Stanton, reuniting them there with the Mescaleros, on the reservation of the latter. Whether this experiment will "work" remains to be seen, as more than half the tribe were dissatisfied, and avowed their intention of returning in the following spring.

Amargo cañon, which is always pretty, and sometimes approaches grandeur, extends westward to Juanita. There it widens out and disappears in a series of little parks, where the mountains diminish into pine-clad hills. For the next score of miles we skirt the turbulent Rio San Juan; but just west of Arboles, where it receives the Rio de las Piedras, we leave it, the road making a long detour, and climbing up and away from the stream, to a wide, rolling mesa. Descending again, La Boca is reached, where we cross the Rio de los Pinos, clear, rapid, and of good size, which we follow up to Ignacio.

At this point is another Indian Agency,—that for the Southern Utes, under an aged head-chief after whom the station is named. There are somewhat over eight hundred Indians here, divided into three or four bands under sub-chiefs. Their reservation, which the railway traverses from where it re-enters Colorado, near Carracas, nearly to the Rio Florida, measures about sixteen miles north and south, and over one hundred miles east and west. These Utes are considered far more intelligent than the Apaches, and their conduct is more taciturn and dignified. Though not congregating in any considerable numbers along the track, they are not unfriendly to the whites, and daily wander about the streets of Durango. They are now the only Indians occupying a reservation within the limits of Colorado.

The members of both tribes are allowed to ride free at will on passenger trains, and the railway company has never experienced the slightest trouble from them. Liquor is kept from their reach as much as possible. Gambling is their passion.

Approaching Ignacio the train runs through shady lowlands, and passes, here and there, groups of teepees, the swarthy occupants of each lodge stepping out and standing motionless as statues in the shrubbery, watching us sweep by. The Rio Florida, which is soon crossed, is alive with trout, and along its upper course is excellent shooting. The whole region is undulating, green-carpeted, and covered with large yellow-boled pines, through which we catch magnificent mountain-views northward. Near Carboneria, the track describes two tremendous loops, in getting down from the table-lands to the valley, and presently, rounding the mountain spur, reaches the Rio de las Animas, which it parallels into Durango, along a cutting through gravel and rock some distance above the bed of the stream.

Toward the last we had seen evidences of the great La Plata coal-field, to which I must devote a paragraph. It extends from the Rio de los Pinos almost to the southwestern corner of Colorado, and has been tapped in many places. This field is in sandstones and shales of the cretaceous age, divided into the upper and lower measures, about 1,000 feet apart. The lower coal measure is in a zone of shaly sandstones which are about 300 feet thick, and when separated from the shale is of excellent quality for domestic use. This lower measure is underlaid by a bed of dark gray shale, containing calcareous seams and nodules, called septaria. The La Plata coal-bed reaches from the east end of the county for over sixty miles, and is crossed by the river. The thickness of the entire bed between the floor and the roof is over fifty feet, and it contains about forty feet of good coal, free from shale. The floor, of grayish white sandstone, is covered with a thin layer of clay and clay shale. Upon this is a layer of compact, firm coal, six to eight feet thick; then a layer of tough black shale, one and a half to two feet thick. The remainder is a bed of excellent coal with only small seams of shale at intervals of four to ten feet. The "roof" is a tough shaly sandstone, alternating with true shales for a distance of several hundred feet above the coal-bed, and containing two or three small veins of coal.

Durango is beautifully located on the eastern bank of the river, the commercial portion being on the first or lower bench, and the residences on the second or higher plateau. Thus the homes of the people occupy a sightly position, apart from the turmoil of traffic, while lofty mountains and wall-like cliffs shelter the valley on all sides. Though founded only in the autumn of 1880, the city now contains a population of over five thousand, and is the most important point in southern Colorado. Here centers the business whose operations extend throughout the entire mountain system, and into the tillage and stock-raising districts of northwestern New Mexico. The great supply stores, with their heavy assortments of general merchandise, indicate a jobbing trade of no mean dimensions, and one which is steadily growing; while the extensive and elegant retail shops, unsurpassed in the state outside of Denver, bear evidence to the refined demands and prosperity of the citizens. Here also are concentrated the social, religious and school advantages which make up an intellectual nucleus. Its low altitude and easy accessibility render the town desirable as a temporary home for those engaged in mining, but who care not to endure the rigors of the long winter among boreal fastnesses. The banks of Durango are substantial institutions, and the hotels are commodious. Municipal improvements are being judiciously added, the most prominent for 1883 having been the erection of water-works, while street cars and gas-works are contemplated at an early day. The smelting of ores is carried on here actively and successfully, the convenience of coal, coke and fluxes, and the hauling of the ores down hill, giving the place marked advantages for this industry. Superior opportunities are likewise presented for a great variety of manufactures, foremost among them being iron and steel productions,—iron ore, limestone and all other necessary ingredients abounding in the locality, and being of easy access. The fall of the stream,—two hundred feet per mile,—supplies a water-power of never-failing volume. Of late the city has been extending its limits, and now one may find an attractive ward, with cosy cottages and more pretentious houses across the river, and in the twilight shadow of the majestic bluffs which here rise precipitously a thousand feet. Taken all in all, no frontier town within our ken shows a more vigorous and healthy growth, or brighter promise for the future, than Durango on the Animas.

XII

THE QUEEN OF THE CAÑONS.

> Receding now, the dying numbers ring
> Fainter and fainter down the rugged dell,
> And now the mountain breezes scarcely bring
> A wandering witch note of the distant spell,—
> And now, 't is silent all—Enchantress, fare-thee-well.
> —WALTER SCOTT.

WHEN, some ten years ago, the writer had let his mule down into Baker's Park, by hitching its wiry tail around successive snubbing-posts, the prediction was ventured that at some distant day a railway would penetrate these solitudes; and that it would approach from the southward, through a cañon which not even an Indian had ever been known to traverse,—the trails in that direction then leading over a terrible range, at a height far above the limit of vegetation. The prophecy has been verified, for the Denver and Rio Grande has already pushed its southwestern extension through the Cañon of the Animas, reaching Silverton in July, 1882.

Here the cores of the Rocky Mountains have been buried beneath an overflow of eruptive rock spreading over four thousand five hundred square miles of territory; or else, along with the sandstones and slates which were deposited against their sides, they have been metamorphosed into schists and quartzites. "The character of the volcanic rocks throughout the district," says Dr. Hayden's report, "is one of extreme interest, demonstrating an enormous amount of activity during a probably short period of time (geologically speaking), which activity was, nevertheless, accompanied by a comparatively large number of changes in the chemical and physical qualities of the ejected material."

This geological composition gives to these mountains,—and particularly to the quartzite peaks along the southern border of the eruptive area,—a different appearance from any of the northern Rockies,—a more precipitous, Alpine and grander countenance, with sharp pinnacles, tremendous vertically walled chasms, and extensive forests of spruce clothing their lower declivities. In no other locality are so many very lofty summits to be seen crowded together. Sierra Blanca and two or three other single peaks in Colorado and Wyoming slightly outrank any here; but nowhere else can be found whole groups of mountains holding their heads up to fourteen thousand feet, and having great valleys almost at timber-line.

The old maps bear the name Sierra Madre, to designate these heights, whose snowy crests filled the northern horizon and forbade the advance of Spanish exploration. The word admits of various applications, but one which might well have been in the mind of him who first used it, is that this vast highland is the mighty *Mother* of our rivers. From its western slopes flow the rivulets that unite to make the Gunnison and Grand,—one of the forks of the Rio Colorado. Easterly, but on its northern face, bubbles the great spring which forms the very source of the Rio Grande del Norte. Every gulch upon its southern breast feeds the rushing streams that furnish to the Rio San Juan all the water it gets for its long journey through the wilderness.

Silverton is forty-five miles due north from Durango; and after leaving the latter point the road leads straight up the Animas valley, here broad and fertile, with green rounded hills sweeping up on each side. Now and then these exchange their softly curving outlines for a bluff-like form, exposing long vari-colored strata of cretaceous sandstones, unbroken, but inclined upward toward the north, where their beds have been gently lifted by a slow upheaval of the mountains. There is much color in this part of the landscape, especially now, when the rains of August have put a spring-like freshness of tint upon everything verdant. The low, treeless benches between the track and the foot of the hills, the open places beside the river, and the pasture-lands are all glorious in a dense mass of sun-flowers, which stand knee-high, with blossoms scarcely larger than a dollar. Thus the outlines of the ridges running in endless succession down to the water's edge, are defined in gilded ranks, that rise behind one another for miles as you proceed. The whole foreground is enchromed; and this valley is the veritable home of Clytie.

EVA CLIFF.

A belt of cedars and dense shrubs stands along the base of the mountains; then perhaps a bare steep space of uniform dull green displays the tone of mingled bunch grass and sage-brush; next will appear a wall of red sandstone set at an angle, and contrasting richly in shades varying from dull vermillion to deep maroon, with the ochre-yellow, white or bluish-gray of the rocks surmounting it. Occasionally these capping-stones show themselves in long, well-exposed strata, slanting to the horizon; sometimes here and there they simply crop out in water-worn crags; again they will be lost altogether under the fringing shrubbery that overhangs the low forehead of the bluff. It is fifteen miles before the valley narrows in, and throughout this whole extent of bottom-land the ground is tilled from the river-brink to the stony up-

GARFIELD MEMORIAL

lands on either side, the fall of the water being so great that irrigation is easy. Ranches succeed each other without any waste land between, and I do not know any portion of the Far West (this side of Salt Lake basin) where the farms seem as thrifty or the houses so comfortable and pleasant. Every sort of grain is raised, and the yield to an acre is large, as must always be the case where the soil is rich, the weather uniform, and the ranchman able to control his water-supply and apply it as he sees need. Garden-produce is much attended to, also, for there is more profit in it than even in grain. Hay and its substitutes, alfalfa and lucerne, take high rank in the list, and of the two last named it is customary to cut three crops annually. In the winter of 1880-81 baled hay was worth $120 and $140 a ton in Durango, while one man told me that it cost him almost $500 a ton to get a supply to his mine in an emergency. In those days the farmer had as good a mine as any on the sources of his river. Such prices will probably never prevail again, now that the railway brings hay and feed from Kansas; but the resident producer can still compete with import figures at a handsome profit.

Two or three miles above Durango we pass Animas City, a small village of unpainted houses, which had an existence and an exciting history long years before its prosperous neighbor was dreamed of; and six miles farther come upon Trimble Springs, directly at the foot of the high bank which here confronts the western side of the valley. It is a singular coincidence, perhaps, that within easy distance and access of all the larger towns or population centers in Colorado, mineral springs are found, whose virtues are sufficiently marked to warrant development, thus supplying each neighborhood its own sanitary as well as pleasure resort. Trimble Springs occupies this relation to both Durango and Silverton, and is greatly frequented by the dwellers in these towns, besides numerous visitors from more remote points. A capacious hotel, of attractive exterior, and admirably arranged and furnished within, affords the comforts of a home. Near by is the bath-house one hundred feet in length, and equipped in the most approved modern style, with all varieties of baths. The temperature of the water as it comes from the ground is 126° F., and iron, soda and magnesia are the predominating qualities, in the order named, while there is also much free carbolic acid gas. The record of cures effected here contains many cases of rheumatism, liver and kidney complaints, and chronic blood and skin diseases; while it is averred that the use of the waters will entirely eradicate the tobacco habit. The temperature is equable, and the surroundings romantic. The river supplies excellent trout-fishing, and the hunter will find an abundance of game in the adjacent foothills. The place must grow in popularity, as it becomes more widely known; for, as the Madame declared, "it excels the White Mountains in scenic features, not to mention the superiority of its thermal founts, and the charm of its climate, over any eastern sanitarium." We marveled at the stateliness of her phrases, but couldn't dispute the facts.

Just at the head of the farming lands, stands the little settlement of Hermosa. I had been there once before this more auspicious advent, after two days of dreadfully weary travel over a mountain trail, and had come down into the valley only to find our much-doubted warnings verified, and these cabins all deserted. We knew what it meant, but made haste to feast upon the green corn, and tomatoes, melons and roots of every sort, which the panic-stricken ranchmen had left behind. Stuffing every available bag and pocket full, we went on to a camping-spot, and deliberated while we cooked our princely dinner.

It was certain that Indians had driven these settlers away, yet there were no signs of hostility apparent. There were five of us, and we had proposed going two hundred miles directly into the Indian country. Should we proceed, or turn back and abandon our exploration? Perhaps if we had possessed only our customary bacon and beans we might have halted; but the luscious corn and melons turned the scale, and we resolved to go forward. Had we not done so we should have missed the rare satisfaction of being the first to tell the story of the Cliff-Dwellers of the Mancos and McElmo. In the nine years which since have worn their footprints into the trail of events, little change had come to this particular spot; and I was glad of it, for it left in my memory a landmark which was lost elsewhere under the obliterating hand of an eager civilization, that has tamed the primitive wildness we rode over in 1874.

Above Hermosa, the valley contracts rapidly, and the wide fields give place to groves of pine, free of underbrush, through which are caught glimpses of the bright stream sinking away from us on the right. The railway commences to ascend the western hills, carving its way along their face, and tracing their shallow undulations by sweeping curves. In places the sharp stones blasted from the roadbed cover the steep and forbidding descent for hundreds of feet below us. Now the river has disappeared, though a rocky ledge marks its cañon confines, the intervening space is wild and broken, and the pines are denser, with great blackened trunks. Presently we emerge into a tiny park, and Rockwood is reached. The location is secluded yet picturesque. Lofty cliffs and precipitous mountains hem it in on all sides, and the meadows in the small depression beside the town are fringed with trees, which are tall and imposing, and yet look more like dwarfed bushes against the massive background of towering bluffs. A lively village has grown up here, whose principal stimulus exists in the fact that it is the forwarding point for the extensive mining district lying between the La Plata and San Miguel ranges. Rico, the most important camp in that section, is connected with Rockwood by a good road, thirty-two miles in length, over which stages and supply-trains make daily trips.

Before leaving Rockwood the train-men are observed to examine critically the wheels, trucks and couplings of our cars, and we know that something unusual ahead suggests the precaution.

Moving slowly through a deeply shaded cutting, a sharp outward curve is rounded, and what a vision greets our astonished eyes! The most magnificent of all the cañons of the Rockies! The mountain presents a red granite front, perpendicular for nearly a thousand feet, and midway between top and bottom has been chiseled from the solid rock a long balcony or shelf, just wide enough for the track. From far below comes to our ears the roar of driven waters, and with bated breath we gaze fearfully over the edge, so perilously near, down, down to where a bright green torrent urges its impatient way between walls whose jetty hue no sun-ray relieves. Overhead the beetling precipice towers ominously, as if about to crush the pigmies who had dared to invade its storm-swept breast. In its shadow all is silent, weird and awful.

The opposite side of the cañon, scarcely the toss of a pebble away, rises almost vertically, a smooth, unscalable wall, that gleams like brightly polished bronze, but is striped with upright lines of shadow, so that it recalls Scott's picture of Melrose Abbey under the harvest-moon:—

> " And buttress and buttress alternately
> Seemed carved in ebon and ivory."

Higher up, the wall breaks away into receding hills, on whose grassy and wooded slopes the sunshine plays hide and seek. A little above the gorge we can discern where the track turns to the right and crosses on a long, low trestle, the alcove in the cañon, while in the loftier heights beyond, the verdure-clad mountains are seen rising into shapely cones and coquetting with the fleecy clouds. Such were the elements of the sublime view in the Cañon of the Rio de las Animas Perdidas caught and perpetuated by our Artist.

Beyond the opening the defile again closes into so narrow a compass that the pines and spruces clinging precariously to the cliffs mingle in a dim arch that spans the chasm. Again the train is creeping cautiously along a dizzy brink, while an hundred feet below the pent-up flood is forcing its passage through the unworn and pitilessly hard rocks. The water is still green as emerald, and has the same luminous quiver and transparence of verdancy which the gem possesses. What gives it that vivid color here in this dark recess?—anything but the fact that it is surcharged with the air caught in its turbulence? We can see great nebulæ of submerged bubbles racing by, meteor-like, too swiftly to rise at once to the glassy surface. Niagara, below the Falls, has that same wonderful, deep green tint. Imprison Niagara, or only so much of it as you could span with a stone's throw; contract its upright, volcanic walls into a crevice sixty feet wide—turn the river up on edge, as it were—and send it down that black, resounding flume, with all the impetus of a twenty-mile race,—then you have an image of this "River of Lost Souls," in the wildest portion of its marvelous channel.

The building of the railway, for the first mile north of Rockwood, exceeded in its daring any work even in the famous Grand Cañon of the Arkansas. The engineer who had charge of the construction showed the Madame a picture one of his surveyors drew of the manner in which the location was made. Evidently the draughtsman took his observations from the water's edge, where his vista was between two walls of natural masonry, and was limited by the side of the gorge which bent sharply there. This wall was vertical and smooth, for almost a thousand feet from its base. From that height were seen hanging spider-web-like ropes, down which men, seeming not much larger than ants, were slowly descending, while others (perched upon narrow shelves in the face of the cliffs, or in trifling niches from which their only egress was by the dangling ropes), sighted through their theodolites from one ledge to the other, and directed where to place the dabs of paint indicating the intended roadbed. Similarly suspended, the workmen followed the engineers, drilling holes for blasting, and tumbling down loose fragments, until they had won a foothold for working in a less extraordinary manner. Ten months of steady labor were spent on this cañon-cutting,—months of work on the brink of yawning abysses and in the midst of falling rocks, yet not one serious accident occurred. "Often it seemed as though another hair's distance or straw's weight would have sent me headlong over the edge," said the chief engineer, and no doubt all his subordinates could say the same. The expense attending such construction was of necessity great, the outlay for this **single** mile aggregating about $140,000.

Crossing the handsome bridge shown in our sketch, the course of the road thereafter is generally on the eastern bank of the stream, although it is recrossed a few times where, by this expedient, expensive excavation **could be avoided.** The water gradually rises to the level of the track, which is henceforth rarely a rod above it. Often in making a curve, one obtains a charming view up the river, with its gracefully **drooping** borders of willows and aspens. Everywhere the mountains **are close at** hand on either side, and a goat could scarcely climb their inaccessible steeps.

Presently a halt is **made, and as we alight, such a** picture is presented as it may never **be our** fortune to again behold. The cañon is compressed into a narrow fissure among mountains of supreme height, whose fronts are in unbroken shadow. At the right a waterfall comes leaping down, to join the foam-flecked river. In the foreground great banks of moss sustain gay flowers, while over **them nod** the stately pines, with swaying vines, keeping time to the fretful murmur of the water. Between and far beyond the clear-cut sky lines of nearer peaks, The Needles lift their splintered pinnacles into the regions of perpetual snow, wrapped in the gauze of a wondrous atmosphere, and their crests glowing as with a golden crown.

Continuing northward, we speedily enter Elk Park, a little valley in **the midst of the range,** with sunlit **meadows and groups of giant pines. As we turn from the** park, a backward **glance** discloses the **subject of our frontispiece,** — Garfield Peak, — **lifting its** symmetrical **summit a mile** above **the track, a peerless landmark among** its fellows.

Onward, the everlasting hills are marshalled, and among them for miles the cañon maintains its grandeur. Frequent cascades, glistening like burnished silver in the sunlight, leap from crag to crag for a thousand **feet down** the mountain sides, to lose themselves in the Animas. **Thus grandly ends this glorious ride, and we** sweep out into a green **park, and are at Silverton, in the heart of Silver San Juan.**

XIII

SILVER SAN JUAN.

*The height, the space, the gloom, the glory .
A mount of marble, a hundred spires!*

—TENNYSON.

N introducing some account of the southern side of the San Juan mountains, as a district producing precious metals, it may be said, in the first place, that it is a section in which productive mining has only very lately been prosecuted in earnest. Its prospects are well-founded ; but almost up to the present time, its inaccessibility and other disadvantages have been obstacles to a development that, under more favorable conditions, would doubtless have occurred. The scrutiny to which it has been subjected by sharp and knowing eyes, and such digging as has been done,— by no means a small amount in the aggregate,— exhibit the fact that the region is remarkable for its general richness. That is, profitable ores are to be had nearly everywhere within its limits ; hardly a hill can be mentioned where veins carrying mineral do not abound. Every square mile of its fifty miles square may safely be assumed to hold one or more good mines It is

NEAR THE PINOS-CHAMA SUMMIT.

doubtful whether anywhere else in the world there is so large a territory over which the most valuable metals are so generally diffused.

"Geologically," we are told on high authority, "the veins of the district are very young, probably having been formed at the close of the cretaceous or the beginning of the tertiary period. The enormous eruptions of the trachytic lava cover a continuous area of more than five thousand square miles. Stress has been laid upon the impregnation with mineral matter of certain volcanic strata,—a phenomenon that occurs throughout a large tract of country. This shows that at the time of the eruptions such conditions existed as were favorable to the formation of that class of minerals generally termed *ores*. It is furthermore to be observed that these impregnations occur mainly in the younger strata. Although the inference can not be drawn that the fissures were formed at the same time, or shortly after the deposition of the trachytic lava, it is allowable to assume that at such a period the material for filling these fissures was existing near the locality where but lately so thorough an impregnation had taken place. The fact that the fissures extend at a number of points, downward, through the older metamorphic rocks, makes it improbable that they should have been formed by contraction of the cooling masses. Singular as it may seem, these lodes are devoid of that which is usually classed as *surface-ore*. Immediately from the surface the perfectly fresh minerals are taken out. The gangue is hard and solid. An exception is made, of course, although only to a slight extent, by pyrite, which decomposes very readily when exposed to the action of atmospheric influences. This characteristic may be explained in various ways,—by the rapid decomposition and breaking off of the wall-rocks, carrying with them portions of the gangue and ore; by the less intense effect of atmospheric agencies; by the character of the minerals composing the ore, and by the comparatively short time that these fissures have been filled. The latter view is the one that would appear as the most acceptable.

"A difficult question arises, when a decision is to be made, as to the causes that have produced the formation of the fissures that were afterward filled. Accepting the theory that volcanic or plutonic earthquakes have probably produced the larger number of all lode systems,—and such we have in this case, it will be necessary to find whence came the requisite force. Along the highest portion of the quartzite mountains we have an anticlinal axis which can be traced westward for nearly forty miles, an upheaval that must have a very perceptible effect on regions adjoining. The idea at first presented itself that this might have given rise to the formation of the fissures, but evidence subsequently discovered demonstrates that long before the eruption of the trachyte, this disturbance had occurred.

"About twenty miles west from the center of the mining region is a series of isolated groups of volcanic peaks. The highest one of these,

Mount Wilson, reaches an elevation of 14,285 feet above sea level, or about 5,000 feet above the valley. Lithologically these groups must be considered younger than the lode-bearing rock of the Animas, and must therefore have become eruptive later. It seems quite possible that the disturbance produced by these eruptions may have resulted in the formation of the present fissures, which subsequently were filled from that source which supplied so much mineral matter to other neighboring rocks in the form of impregnation."

This ore, then, may be set down as principally galena,—a **lead ore of silver, frequently enriched by gray** copper (tetrahedrite). **The high percentage of lead makes** smelting the most rational process of treatment, and they are generally to be classified as smelting ores.

In several localities, however, of which Parrott City and Mount Sneffels are chief examples, rich ores of silver are found, nearly or quite devoid of lead. These come mainly into the group of antimonial ores, with chlorides and sulphides also. Popularly these ores,— barring the chloride,— are termed "brittle silver," and on account of the absence of lead, they are unfit for smelting, but must eventually be treated by a milling process in which the pulp is subjected to the action of mercury in amalgamating pans, where the silver is separated from the quartz and collected by the quicksilver. Antimonial ores, prior to amalgamation, will require chlorination, that is, roasting with salt, as is done at the Ontario mine, Utah ; while the chlorides and sulphides of silver can be treated directly, without roasting, as at the mines of the Comstock lode, Nevada.

The foregoing remarks apply generally to all of the mining districts mentioned in the present chapter ; and their uniform nature is readily explained by the fact that the whole neighborhood is of the same geological age, character and origin.

The mines in the immediate vicinity of Silverton, my starting point, are situated upon, or rather *in*, the lofty mountains which hem in the little park. Southward of the town, easily recognized by its cloven peak, stands the Sultan, thirteen thousand five hundred feet in altitude. Its most noteworthy mines are the "North Star," "Empire," "Jennie Parker," and "Belcher." Tower and Round mountains, next northward, contain several ledges of low-grade galena ores of silver.

Crossing the Animas to the eastern side, King Solomon wears as the central jewel in his crown another "North Star." It stands upon his very brow,—one of the loftiest silver deposits in the world, almost fourteen thousand feet above the restless surf of the Pacific. Here, too, the ore is galena and gray copper of extraordinarily high grade. A marvelous trail has been cut through the woods and then nicked into the almost solid rock of the bald mountain-crest, far above timber-line, or built out upon balconies of logs, along which burros carry to the mine all its supplies, and bring down its product. On King Solomon are several other noteworthy claims, such as the "Shenandoah," "Eclipse," and "Royal Tiger."

Nothing but a bird or a mountain sheep would be likely to attempt the almost vertical wall rising from the southern side of Arastra gulch to culminate in the spires of Hazelton mountain. Coming out into the valley, however, a road is to be found zigzagging its way up the slope leading to the principal mines, pierced only a trifle below the border of stunted spruce-woods. Very likely Dr. Holland was correct in his poetico-mineralogical statement that

> "Gold-flakes gleam in dim defiles
> And lonely gorges;"

but it is certain that in the San Juan, silver resides upon the loftiest ledges, where the shadowy peaks form "bridal of the earth and sky."

The group of mines to which I have referred are known as the "Aspen," consolidating several names of properties under the ownership of the San Juan and New York Mining and Smelting company, which is also proprietor of the smelter at Durango.

Sitting in a cozy office one evening, with two or three pleasant visitors, the conversation fell upon the other side of the year, for the last man came in rubbing his knuckles as though it were cold.

"Ha, ha!" laughed the merry Madame, glancing out at the ashy-gray peaks, which were wan in the new moonlight with autumn's first white dusting; and, as she laughed, she quoted

> "Once he sang of summer,
> Nothing but the summer;
> Now he sings of winter,
> Of winter dark and drear;
> Just because a snow-flake
> Has fallen on his forehead,
> He must go and fancy
> 'T is winter all the year."

"Well, it *is*, pretty nearly," comes the quick rejoinder. "I have seen it snow every day during the last week of August, and the seasons which do not give us frosts in July are rare. When I first came to Baker's park I asked a miner what sort of a climate reigned here. 'It's nine months winter and three months mighty late in the fall!' was his laconic report, and I have found it a true one.

"You see," he continued, "winter really begins about the first of November. The superintendent who hasn't got his supplies at his mine-house by that time had better hurry, for some morning a storm will begin which will drop three or four feet of snow on a level, and fill all the small gulches full. Then his chance of packing anything up the mountain is gone. In 1880, several foremen were surprised in that way, but the first storm came remarkably early,— the 8th of October,— and on the 11th the snow was five feet deep. Later there was an open spell, though, when deficiencies could be made up, but that was only luck."

"But," said the Madame, solicitously, "how can men live in those little cabins, away up there, all through the terrible winter? I should think they would freeze, or that avalanches would sweep them away."

"Oh, both those misfortunes can be guarded against. The houses are very tight and snug, and fuel is carefully stored away. Then, too, the work is carried on underground, where the temperature is practically changeless the year 'round, and the men have little occasion to go out of doors unless they wish to, for the entrance to the mine is under the shelter of the house-roof. Then, too, the fact is, that bleak and thoroughly arctic as it looks, the mercury will not fall so low, or at least will not average as low, up at timber-line on Sultan, as it will down here in town. I suppose the excess of dampness in the valley makes the difference, which is more apparent to our feelings than even to the thermometer."

"But the snow-slides are **sometimes** terrific, are they not?" is asked.

"Terrific? I assure you that word is not half strong enough to express it. When you go up to Cunningham gulch, and over into the other valleys, you will see the sides of the mountains, in certain places, utterly bare of trees two or three thousand feet below the limit of their growth. That is **where they** have been swept away and kept down by constantly recurring avalanches of snow, which in many parts of these ranges are liable to slip down in masses perhaps a mile square and anywhere from ten to a hundred **feet** deep, bringing **rocks and everything else with** them. Of course, no sapling could stand such a scraping,— nothing can. I was in a slide once, and I can appreciate it, I assure you."

"You were!" exclaims the Madame, round-eyed at this. "I thought you said nothing could stand a snow-slide."

"I didn't attempt to. I went with it, and was carried down the mountain-side head first,— most of the time under flying clouds of snow-dust, until I plunged—fortunately feet down—into the compact mass at the bottom. Then a friend followed and dug me out, happening, by good luck, to begin his prospecting in just the right place."

"But weren't you smothered; **and** how did **you feel** going down?"

"Very nearly smothered; as to the feeling, it was merely a confused sense of noise, darkness, nothingness, nowhereness, and the sudden end of all things. Can you understand such a combination of sensations?

"Here in the park," our friend continued, "we don't mind the winter much. We have enough people to keep one another company, and we have no end of fun snowshoeing."

"What sort of snow-shoes?" I break the silence to ask.

"The Norwegian *skidors*,—thin boards, ten or twelve feet long, and slightly turned up in front. There is an arrangement of straps about a third of the way back from the front end, and that's all there is of it, though it's a good deal if you don't know how to manipulate—"

"*Ped*ipulate, would hit it closer, wouldn't it?"

"Suit yourself; you won't choose your language so carefully when you suddenly find yourself filled full of snow, after an involuntary header, and one or both of your snowshoes going on down the moun-

CHIEFS OF THE SOUTHERN UTES.

tain like a race-horse. If you stay here a winter, though, you must learn, unless you are willing to remain cooped up in your cabin from November till May. There is no other possible way of getting about. Before the railway came, all our mail was snowshoed in, and it was very likely to be delayed two or three weeks at a time. Then we have debating societies. 'Resolved, That a burro has no rights a miner is bound to respect,' was our first question one winter—and parties and balls. Formerly, in the older communities, merchants could easily calculate the extent of their sales during the cold months, but those in the new camps, the first winter, sometimes saw hard times.

"We came very near a famine in Rico, the first winter," our visitor continues. "Nobody could tell just how many people would stay, the winter closed in unexpectedly early, and, all together, before New Year's day, it began to be whispered that the supply of 'grub' was short. As fast as the stores diminished, prices went up, until they were nearly fabulous. Everybody was on short rations alike. The hotel would give beds, but no board. One day a miner came in from over the range on snowshoes, and reached the hotel nearly dead with hunger and exhaustion. Pfeiffer took pity on him, as an exceptional case. 'I gifs you your supper,' he said, 'und a ped, und I gifs you one meal to-morrow; after that you must rustle for yourself.' Flour, bacon, ham, sugar, coffee, everything, even tobacco, gave out in the shops; and had it not been that one of the mines which had laid in a large stock of food, shut down and so sold out, it is probable that the whole camp would have been obliged to have dragged themselves through the depth of a hundred miles of February snows, out into the lower country. Comical stories are told of how the first burro-train load of provisions was distributed."

But in spite of all this isolation, this necessity for elaborate preparation, the arctic altitude and polar length of the "season of snows and sins," as Swinburne phrases it, the winter is really the best time in which to work these silver mines, and the impression that the San Juan district must be abandoned for half the year, is entirely wrong, when any thorough system of operations has been projected. Well sheltered and abundantly fed, removed from the temptations of the bar-room, which can only be got at by a frightfully fatiguing and perilous trip on snow-shoes, and settled to the fact that a whole winter's work lies ahead, there is no season when such steady progress is possible, either in "dead-work" development or in taking out ore preparatory to shipment in the spring.

Two little streams come down to the Animas at Silverton—Mineral creek and Cement creek—the former passing between Sultan and The Anvil, and the latter between The Anvil and Tower mountains. Up Mineral creek a dozen miles we find Red mountain, the scene of the latest and richest discoveries in the San Juan, but which will be considered elsewhere. Cement creek has several good mines, while beyond, almost on the divide between the Animas and the Uncompahgre, lies the Poughkeepsie Gulch camp, which was, not long ago, the locality of a "boom;" and I have the opinion of a very competent judge, that there probably is no equally limited district in the whole region, Red mountain being perhaps excepted, where so much good ore exists.

Still farther, at the very head of Cement creek, is located the important Ross Basin group of mines, worked by English capital, as are many other claims in the San Juan mountains. A neighboring mine is remarkable for producing an ore of bismuth in such quantity as to give it great mineralogical interest. Bismuth is exceedingly rare. In the United States it is obtained only to a small amount in Connecticut. Saxony furnishes commerce its main supply, procuring it at the metallurgical works of Freiburg, where it is associated with the lead ores, and is extracted from the cupel furnace after large quantities of lead have been refined, being accumulated in the rich litharge, or liquid dross, near the conclusion of the process. This litharge is treated with acid, and the bismuth precipitated as a chloride by dilution with water. The making of lily white and other complexion compounds is the chief use to which bismuth is applied. The Madame assures me that the effect upon the skin is very noxious,—but how could she know that?

Crossing the divide, passing the "Mountain Queen" district, and proceeding eastward down the west branch of the Animas to the town of Animas Forks, another prosperous and populous mining area is reached. Mineral Point, where twenty or thirty rich veins crop out, is covered with claim stakes until it looks like a young vineyard. Its ores, in general, are dry,—that is, contain little lead; and some streaks show the beautiful ruby silver. Yet further down, on the eastern side of the river, lie the partly developed silver veins and ledges of gold quartz in Picayune gulch, where hydraulic machinery is used; and opposite is Brown's gulch, where galena and gray copper occur.

The river in this part of the valley struggles through a close and pretty cañon, at the lower end of which stands Eureka,—a neat village nestling among trees. Here, too, are concentration works, and the headquarters of several companies operating in Eureka, Minnie, and Maggie gulches.

The wall like sides of the mountains shutting close in together from Eureka down to Howardsville, show "mineral stains" everywhere, and the eye can trace dozens of veins slanting up and down the dark cliffs, and study how they thicken here and pinch there, or just beyond perhaps disappear altogether, upsetting all the old theories. At Howardsville, which was the center of everything years ago, the reviewer diverges up Cunningham gulch, completing the circle of his inquiries, for from Howardsville to Silverton it is only four miles.

Cunningham is a good type of those huge ravines the western man calls gulches. Its real walls are several hundred yards apart,—Galena and Green mountains on the north, King Solomon on the south—but from each have tumbled long sloping banks of débris, that join at their bases into a series of ridges. Among these a turbulent stream seeks its irregular way, and over them the traveler must climb wearily, making frequent detours to avoid huge pieces of rock that have fallen bodily from the cliffs, and have been rolled by their great weight to the very bottom of the gorge. Here and there the walls are sundered, and down a side ravine is tossed a foaming line of cataracts; or some hollow among the peaks (themselves out of sight) will turn its gathered drainage over the cliff, to fall two or three thousand feet in a resounding series of cascades, white and filmy and brilliant against its dark and glistening background. Wherever any soil has been able to gather upon these loose rocks, if some curvature of the cliffs protects from the sweeping destruction of snow avalanches, heavy spruce timber grows, and this, with lighter tinted patches of poplars, or willow-thickets in wet places, or a tangle of briers hiding the sharp rocks and beloved of the woodchucks and conies, give all there is of vegetation.

But these are all minor features, under-foot. Overhead tower the rosy and gleaming **monuments of that old time** "when the gods were young and the world was new;"—cliffs rising so steeply that only here and there can they be climbed, and **studded with domes** and pinnacles so slender and lofty that, under our unsteady glance, **they** seem to **totter** and swim vaguely through the azure concave.

Amid this magnificence of rock-work, spanned by a violet edged vault which is not sky but only color,—the purest mass of color in the universe,—passes the trail and stage road cut over the lofty crest to the sources of the Rio Grande, and thence down through Antelope park to beautiful Wagon-Wheel Gap, and the railway again. Here, too, are rich silver mines, lowest down the "Pride of the West," next the "Green Mountain," and, last of all, "Highland Mary," standing almost on the summit of the pass.

The central point and outlet of all this district is Silverton, and its founders preëmpted almost the only site for a town of any consequence in the whole region. Yet she has less than one thousand acres to spread herself over. Engulfed amid lofty peaks, a little park lies as level as a billiard-table, and as green, breaking into bluffs and benches northward where the river finds its way down.

When, three and twenty years ago, miners· were amazed at the wealth disclosed in the mines of central Colorado, eager prospectors began to penetrate yet deeper into the recesses of the jumbled ranges that lay behind the front rank. Among the boldest of these was a certain Colonel Baker, reputed to have got his title as a confederate officer, who organized a large party of men—some say two hundred in number—to go on an exploration of what was then called the Pike's Peak belt, including nearly all the region between that historic mountain and the head of the Gila river in Arizona. Marching eastward to Pueblo, and thence by the old Mexican wagon-road through Conejos and Tierra Amarilla, Baker and his men worked northward along the San Juan and Animas, prospecting for and finding more or less bars of gold-gravel (you may get "colors" anywhere in this part of Colorado), till finally, in the summer of 1860, he crossed the range, and discovered the deep-sunken nook which bears his name.

Erecting a central camp here, these prospectors climbed all the mountains, and pushed up every ravine, in search of gold, but found small encouragement. The silver they knew of, but had no means of working. Winter came, and they gathered together and built cabins in the thick timber at the mouth of the cañon. The snow packed deep about them, a provision train intended for their succor was captured by the Indians, who became aggressive, sickness set in, and the horrors of starvation stood at their very doors. This terror, added to their lack of success, overcame even pioneer patience and philosophy. Reviling Baker as a cheat, who had brought them, under false pretenses, into this terrible state, they were about to hang him to one of the groaning pines that mocked their misery with a loud pretense of grief in every storm, when some slight help came and the colonel's neck was saved. The following summer, all who had not died crawled out of their prison-park and returned to civilization.

It was not until ten years later that any persons went to Baker's park to stay, and then they were extremely few in number. Almost the first result of this second advent of prospectors was the unearthing of the "Little Giant" gold mine in Arastra gulch—a narrow ravine where were at once erected a log village and an arastra with which to crush the quartz, worked by the little stream which trickles down from the snow-banks. Simultaneously came the discovery of silver leads, a fact that speedily got abroad, induced a little boom, and set Howardsville on its feet as a camp of some importance and magnificent expectations.

Five miles below, Silverton was laid out straight and square, became the county-seat, and attracted most of the new-comers as a place of residence. At first, of course, all the buildings were of logs, and bore roofs of dirt. To-day the village has perhaps fifteen hundred permanent residents; churches, schools, newspapers, the telegraph, and all the appurtenances of frontier civilization. It is characteristic of these mountain towns that they spring full size into both existence and dignity. There is no Topsy-growth at all; rather a Minerva-like maturity from the start.

For several years no wagon-road entered Baker's park, and the only communication between it and the outside world was by saddle animals. As the local paper gently expressed it, it "was somewhat deprived of easy transportation." Goods and machinery of every sort had to be brought in on the backs of the tough and patient little Mexican donkeys, toiling across the terrible heights under burdens almost as bulky as themselves. The whole town would be alive with a general jubilation when the tinkling bells of the first train of jacks was heard in the spring, for that meant the end of a six months' siege in the midst of impassable snow.

Though these mountains are yet full of men who go about all day with a big six-shooter in their belt; and though the main streets of Silverton (like other frontier places) contain too many drinking and

CAÑON OF THE RIO DE LAS ANIMAS.

gambling saloons, yet the town has never passed through such a rough history as most mining camps see, and it is to-day the most orderly village in the whole region. This is chiefly due to the quietly determined attitude its best citizens have taken, and their fixed purpose not to let the lawless rule.

In the summer of 1881, however, the remnants of the gang of desperadoes who had infested Durango during the winter, tried to make Silverton a rendezvous, and one night killed an inoffensive and highly esteemed officer, who was aiding a sheriff to arrest one of their number. It was the culmination of many atrocities, and the citizens at once resuscitated their Vigilance Committee. One of the ruffians was apprehended the same night, and quietly hung the following evening. Large rewards were offered, detectives and sheriffs set at work, and finally the leading spirit of evil was captured by the treachery of his most trusted ally in previous villainies. After some delay this prisoner was brought to Silverton in charge of his Judas like comrade, who took his reward and rode swiftly away, distrusting the pledge of the citizens that he should go safely out of town. This was on Friday. The prisoner was locked up, and strong relays of heavily armed guards, chosen from men of respectability and standing in the community relieved each other at the jail night and day, until Sunday morning came, and with it a cold, dismal storm.

All day the rain fell steadily down, and the air was clammy with chill mist. Dense banks of clouds were packed into the dripping gulches, capped the hidden summits and clung in ragged masses among the trees that darkly clothed the sides of the mountains. Occasional gusts of wind drove the storm hard against the window panes, but for the most part rain fell quietly, the streets became avenues of inky paste, and the darkness of evening gathered early about the town, settling like a pall upon all the waiting people in it.

Everyone knew, though the majority could hardly say why, that the hour of fate had come. As the night thickened, men gathered on the corners nearest the jail, and, unmindful of the persistent rain, stood talking in low tones to two or three listeners whose faces were close together and strangely serious. Moving here and there were other little groups, their footsteps hardly heard in the soft mire, and their voices hushed,—moving chiefly up and down the alley where the jail stood.

The saloons and gambling-rooms were open, but the dance-hall, which last night echoed so late to the clatter of heavy boots and the shouts of half drunken revelry, was closed, and the few women who haunted the other liquor dens seemed to have forgotten their coarse jibes and laid aside their accustomed wiles. The soft rattle of the thin faro-checks, the clink of silver lost and won, and the louder crack of billiard-balls, were heard as usual, only more distinctly, while the monotonous "ante-up, gents!" "Are you all ready?" "The deuce wins," and so on, of the imperturbable dealers, mingled in a sort of minor music to which all sharper sounds were accordantly attuned. But the players were moderate in their stakes, and the ordinary excitement of the smoke-dimmed rooms was hushed.

Still fell the rain drearily. The stern guards about the jail hugged their rifles under their arms, to keep them dry at the breech, and now and then tipped streams of water out of the broad hollow brims of their *sombreros*. In the log gaol the murderer lay upon his couch, apparently sound asleep, and the inside sentinels rested their guns on their knees and counted the moments until their watch should be over. Nine o'clock came and passed without note. Nine o'clock and thirty minutes was marked on the cold face of the clock, when the key grated in the iron lock, the door opened a little way and three masked men glided in, shutting the door behind them. One brought with him a rope, which he fastened into a staple set in one of the rafters, standing upon a chair which gave him only height enough just to reach the beam. Another touched the prisoner, and told him his time had come. That afternoon he had assured his keepers that they would see "as brave a death as ever went out of that prison." It was no surprise, then, to see this boy (for he was scarcely twenty) rise coolly from his bed and walk to where the chair had been placed underneath the dangling noose. Perhaps he would have liked to have shaken hands, had not his arms been manacled behind his back; but instead, pausing a moment ere he took his place, and without a tremor, he simply said, "Well, *adios*, boys!" Then, stepping up, he inclined his head and himself set it well within the noose. There was a touch of the rope to tighten the knot, a snatching aside of the chair, and the outlaw had "gone over the range," beyond all further harm or doing of it.

Then the jail was locked, and few knew, even at midnight, whether or not the retribution had come. There was no boisterousness, no gloating over vengeance satisfied, less of mirth and curiosity, than I ever saw in a community where an execution under the sanction of law was taking place. It was more an awe-struck feeling of a terrible necessity, as if an impending calamity was at hand, or some great affliction present.

Next morning the coroner's jury met, and a ray of light was shot across the sombre picture; the verdict said:

"*Came to his death from hanging 'round!*"

XIV

BEYOND THE RANGES.

All the means of action—
The shapeless masses, the materials—
Lie everywhere about us.

—LONGFELLOW.

THREE districts require mention before this corner of the state is bidden farewell,—Ophir, Rico and the La Plata mountains.

Ophir lies fifteen miles east of Silverton and on the Pacific slope, for it is at one source of the Rio Dolores. It is reached by a wagon-road up Mineral Creek, which is one of the most "scenic routes" I know of in Colorado. At first there is not much to call forth admiration; nearing the top, however, a remarkable picture presents itself. In a closely guarding circle of purplish peaks, stand two isolated mountains of entirely different character and most striking appearance. Instead of the vertical cliffs, serrated and splintered summits and ragged gray of the majority of the mountains, these are as rounded and smooth on top as if they had been shaved by a lawn mower, and rise in unbroken slopes far above the blackish masses of timber which closely envelope their bases. It is their color, however, that makes them so grandly conspicuous. Long strokes of orange and rust color extend up and down from the spruces to the apex, streaked with bright red and set off with upright lines of glowing yellow, all softly blended together and crossed by a crowd of hair-lines, wavy and level with the horizon, like the plumage of a canvas-back duck. Stand where you will on the eastern side of this divide between the Animas and the San Miguel, and these great, smooth, cushiony hills of red, tower up level with your eye, burning under the sunlight.

At last the road rises above timber line, but even to the last verge, the soil under the trees is crowded with flowers and all sorts of pretty herbage, among which the strawberry takes precedence in point of abundance. Then the track lies underneath beetling cliffs which have crumbled into long tall, and the pass itself is only the triangular depression between two opposite slides. On one side here the rock is brown and broken almost as fine as railway ballast; on the other the fragments rule much larger in size, are of bluish trachyte and completely covered everywhere with a stone-lichen hardly thicker than paint, which gives them a decidedly green color, while the brown rocks opposite are entirely devoid of lichens.

Down this jumble of fallen rocks—the scene of one incessant slow avalanche from the weather-crumbled crests still remaining above—the road passes by a steep and tortuous grade, made somewhat smooth by filling the crevices with small stuff ; but the result would make the ghost of McAdam turn a shade paler.

These vast "slides" are a prominent feature in every landscape in southern Colorado. The volcanic rock with which all the mountains are capped, has a natural cleavage in two directions and rapidly disintegrates, even under the air. On the quiet, still days of midsummer, you continually hear the rattle of pieces of rock which have fallen untouched from some scarp or pinnacle, and are racing down the steep talus below. The winter, however, is the time of greatest destruction. Into the thousand cracks and crannies the rains and snows of autumn pour floods of water, which penetrate the inmost recesses of the well-seamed crags. Then comes a frost. The little veins and pockets of water expand with a sudden force, combined and irresistible. Perhaps some huge projection of cliff flies to pieces as though filled full of exploding dynamite ; perhaps a stronger body of frost behind it pries off the whole mass at once, and it dashes head-long down the side of the mountain, to scatter widely its cracked shell and leave the core a huge bowlder, which crashes its way far into the struggling woods at the foot of the rough slope. This process goes on, season after season, until finally the thousands of feet of summit, which once towered proudly above the mountain's base, have been crumbled down to a level with the top of the débris-slope. If the rock is very soft, then the process goes on with each fallen block, until it is reduced to soil and forms a smooth, grassy slope, or a clean shaven but barren slide, like the rich red hills we saw on the other side of Lookout ; but if the fragments are hard, then gradually the bushes and grass will creep up, and the forest will follow as high as climate and snow-fall will let it grow, and above will be a rounded crest of broken lava, like Veta mountain—the worst thing to climb in the wide world.

From the long, slanting niche which lets the road down across this broken and sliding rock, where men are always at work to throw aside the ceaselessly falling crumbs of the cliff, one gets his first view of Ophir gulch,—a valley half a dozen miles in length, without an acre of level ground in the whole of it. This end is closed by Lookout mountain, the opposite by the lofty crags of Mt. Wilson. On the north, Silver mountain cuts the sky in ragged outline, and, braced against its base, Yellow mountain rises straight from the creek-side to an almost equal altitude. In the crevice between stand the score or so of log cabins, which constitute what many persons consider the liveliest camp in the whole San Juan.

It is only eight years since the value of this locality was made known, but now the mountains on both sides of the gulch are pitted like a pepper-box with prospecting tunnels, and there are perhaps twenty mines shipping ore in profitable quantities, even under the great disadvantages of their isolation. The leads in general run northeast and southwest, but good openings have been found all the way from the brink of the creek to the shattered combing that casts its ragged shadow down the long, white slopes. Systematic development has been carried on in very few mines as yet, but the indications promise great things for the future. Half a dozen gold workings in particular are very rich, and several sales have been made exceeding $50,000 for a single location.

Remounting, the ride homeward through the mellow afternoon, was very delightful. The mountains rose on either side high above where the hardiest trees could manage to exist, gorgeously stained in great chevrons of red, orange and rust yellow. Lookout and its brother peaks seemed vast stacks of triangles, all upright and baseless, backed with long slides of varied umber tints. On some of these slides the grass has grown, long tongues of it penetrating far toward the bright walls overhead, while elsewhere mile-wide slopes of grayish white lie untouched by any blemish or projection. Everything is triangular,— the outlines of the peaks and the reverse in the gorges between; the shape of the fallen fragments; of the long spear-points of verdure that climb them, and of the trees and even the separate leaves that blend into those acute green patches; of the broad strokes of vivid color that have been painted so lavishly on these splendid slopes; even of the splitting and cleavage of every cliff-face and toppling spire that glistens in the slanting light and throws a slender-pointed shadow across the velvet brim of the valley.

Backward, where the forests lie unbroken on the southern wall of the gulch, long ranks and patches of aspens were interspersed with the reigning evergreens, and these the frosts had touched with various hues from its full palette—bright green still where the leaves were protected, yellow on the warm side of the ridges, vivid orange and scarlet along the crests,—so that these patches glowed like red and yellow flame against the dark spruces and firs.

Near timber-line there is a remarkable picture. Down from the northern mountain there trickle reddish streamlets over a space several rods in width. A few yards below the road all this water collects itself into a basin, which, begun by some trivial obstruction, has been able to build up its walls by slow deposition, until a great iron tank, with walls twenty-five or thirty feet high, and several feet thick, contains all but a trickling overflow of the mineral water. This tank is surrounded with pretty trees, and its wavy red outline holds a fountain as richly green as an emerald; or blue if you look at it from some one of the surrounding heights, so that the Spanish way of calling a spring *ojo*—an eye—seems very natural.

Beyond this highly tinted natural reservoir, built out like a balcony on the steep hillside, you look across to undulating verdant knolls, where shapely trees are scattered thinly, up beyond a deep maroon slope, falling from a noble, iron brown bluff, and so on away to the gray and lofty peaks, in whose rifts and vertical gorges the shadow lies blue as the farther edge of the sea, and whose clustering, cumulative spires, culminate in gleaming apices of snow.

Rico is the next point. It is accessible from the north by wagon roads, but the entrance from this side is by stage from Rockwood Station on the railway, midway between Durango and Silverton. The road bears northward, and the views to the eastward are far-reaching and noble. The traveler alive to the resources of the region, will note the

ON THE RIVER OF LOST SOULS.

rich, thick grass, and the great pine timber, with poplars between to serve for log house and fencing purposes; he will also regret the limited possibilities for agriculture. Toward the head of this valley the woods thicken, and the road gets rougher and starts up the long slope that ultimately carries it over the hill. The ragged outlines of the San Miguel range come into view ahead, while the valley below, a solid "heather" of scrubby oak bushes, briers, ferns, and so on, seems carpeted in a queer design of tints of green and yellow, interspersed with all the mixtures of orange, scarlet and crimson that the deft fingers of the early frost could devise.

Over the long hill and past the spruces, an hour's trotting takes the buckboard through the long hay meadows of Hermosa park, whence it ascends a four-mile hill to the summit of the last range dividing the waters of the Rio Las Animas from those of the Rio Dolores. And how we rattle down that Dolores slope ! An Englishman riding on the Pennsylvania's sixty miles-an hour train from New York to Philadelphia, the other day, exclaimed, "It's wonderful ! I think if something should drop one of you Yankees astride a thunderbolt, the first thing you would do would be to say, 'chk! chk!' " I thought of that as we started, almost at a gallop, down that steep and winding mountain road. Corners — we snapped around them. Hollows and ridges — we bounced into and out of them. Down long, rough slopes, cut in the side of a hill so steep that just under the hub it fell away hundreds of feet almost like a precipice; down through the full blaze of the afternoon rays in the frost-turned aspens, where

> "Tremulous, floating in air, o'er depths of azure abysses,
> Down through the golden leaves the sun was pouring his splendors,"

we rushed at a pace that Phæton, in his first hours of freedom, might have enjoyed in his chariot, but which to us, in an old buckboard, was simply torture. Why we did n't pitch off the imminent verge, why we did n't fall to pieces against some one of the thousand rocks we assaulted, why our bones were not broken and our diaphragms bursted, is incomprehensible.

Rico is situated in the center of a volcanic upburst which has parted the sandstones and limestones once spread thousands of feet thick over the area, and whose edges now stand as bold bluffs all around this break, which is nearly four miles in breadth and about eight in length. The town itself is made up of a scattered, gardenless collection of log cabins and some frame buildings, with a log suburb called Tenderfoot Town, and numbers about six hundred people. It is very dull, compared with most Colorado camps, but this is owing to the fact that everybody is waiting until the railway gets a little nearer.

The Rico mines are characterized by their great dissimilarity with each other. Nearly every sort of ore, of both silver and gold, is found mingled in a most heterogeneous way among the lavas, recalling that marvelously mixed mineralogical madrigal in the Colorado comic opera, *Brittle Silver.*

> "I have found out a gift for my fair,
>   I have found where the cálcites abound,
> Where sklópsite and zircon appear
>   With sárcolite scattered around.

> "Then come love, and never say nay,
>   With pterosmine thy heart I'll delight,
> With diaspore and mangandblend gay,
>   And pharmakósiderite."

Some true fissure-veins exist, but more irregular deposits, and both "lead" and "dry" ores occur, often in contiguous claims. The richest ores thus far are those without lead ; where galena occurs it is mixed with so much zinc and antimony as to make it troublesome in treatment. A galena ore here, which will show a mill-run of thirty ounces (my authority is Mr. Amos Lane, superintendent of the smelter), is considered very good.

Rico has not yet worked far enough into her very numerous "locations" to make sure of the riches her mountains are supposed to contain. There is no doubt that the cliffs about her are full of silver and gold, stored up in what, under more favorable circumstances, would be profitable quantities ; also that there is in the near neighborhood a magnificent supply of bituminous and "free-burning anthracite" coal, good material for charcoal, limestone for flux, bog and magnetic iron, fireclay and good building stone. The time will come, then, when Rico will be able cheaply to treat its own product, but this will be after wagon-roads and railways have come nearer, and outside capital has lent its strength to bring to the surface the hidden, or only partially exposed, treasures of the veins.

South of the San Juan range, and somewhat isolated, is the noble La Plata group of mountains. They are volcanic, like the rest, and, of course, of Alpine appearance, while their slopes, lying far south, produce so many varieties of foliage, that they often present real bits of beauty—a word having rare application in Colorado's scenery. These mountains were prospected eight or ten years ago, and a placer bar of supposed extraordinary value was found near the head of the Rio La Plata by a company of California miners. I remember very well the picturesque little camp they had there, and the day they got their first butter for nine months. Having interested in the locality Mr. Parrott, a California capitalist, a town grew there rapidly, called Parrott City, now only sixteen miles from Durango, and arrangements were made for working the placers by hydraulic machinery. Meanwhile searching about the peaks disclosed gold quartz in some quantity, and many veins bearing dry ores of silver, absence of galena being characteristic. I see no reason why these peaks should not be equally productive with any district in the region.

But this is true, as I constantly insist, of all the San Juan. Everybody looks forward. Each proposes to do this and that, and to be happy—"when I sell my mine." Perhaps this delicious uncertainty is a part of the fun. Yet many a miner would reprove me for exaggerating the uncertainty ; I only hope he is right and I am wrong. That there is a vast amount of the precious metals hidden in the veins of these mountains is undeniable. It is equally true that we know where very much of it lies. But the question stands : Is it sufficiently concentrated to make the getting it out and refining it into a useful condition, yield a margin of profit on expenses? No doubt it is in many cases, but is it in the majority of so-called "mines," or in enough to support any general population and business? Many discreet persons say "No." Many more, naturally, will answer, "Yes." I, myself, making no claim to utter a skilled, or a weighty, or any kind of an opinion except a carefully unbiased one, think the balance of chances is in favor of ultimate success; and I am not afraid to predict that through slow but permanent advancement this corner of Colorado will come to be one of the most important silver-producing regions on the globe.

Upon this event depends the fate of a great many enterprising investments. Faith in the success of these mines has caused the Denver and Rio Grande to build two hundred and fifty miles of railroad over mountains and wide plains which of themselves would never support the line. Faith in these mineral treasures has caused hundreds of men to follow the railway, and has set on foot little towns all along its track; and a part of the same faith is all that keeps alive the thriving town, Durango, where scores of well packed warehouses vie with one another in plethoras of merchandise, and thousands of men are exciting each other in pushing, plucky struggles after the supremacy of wealth. The miner picks away at his rock, and hopefully pays for his supplies until the last dollar is gone, and then goes at work earning more in the service of his more fortunate companion. The patronage of these men, always just on the brink of a "rich strike," is what keeps this southern Denver—scarcely four years old yet—alive and sturdy. The precious minerals can only be procured in this region by hard and skillful labor; they are not in carbonate-beds or placer-bars, to be picked to pieces and reduced at trifling cost. On the other hand they are richer, and while the profits are no less than in the former case, the expense of getting out is several times greater. This means the disbursement of far more money in the locality for the same amount of value received from the mines by the owners, than in an easier district to work — Leadville, for example. Thus an ore which would yield only sixty dollars to the ton will pay to work, very likely, in a carbonate camp, since it would cost only ten dollars to get it out and through the smelter; while to get the same profit on a ton of San Juan ore, it must carry one hundred dollars to the ton, say, since it requires fifty dollars to mine it. Thus for every ten dollars spent in an easy locality, five times as much must be expended here; or, in other words, five times the population maintained under the former circumstances, will be supported here, and be permanent, for fissure-veins do not produce spasmodic and uneven results, but continuous, progressive and practically inexhaustible supplies of ore for the proprietor, wages for his workman and business for the merchant, artisan and shipper. All this is the best kind of an outlook, and means that the San Juan will always be a good country for the man of moderate means, although the mining speculator may consider it too solid and tangible to suit his purposes, and therefore be loath to praise it.

XV

THE ANTIQUITIES OF THE RIO SAN JUAN.

> Dismantled towers and turrets broken,
>   ' Like grim and war-worn braves who keep
> A silent guard, **with** grief unspoken,
>   **Watch o'er the** graves **by the Hoven weep,**
> The nameless graves of a race forgotten;
>   Whose deeds, whose words, whose fate are one,
> With the mist, long ages past begotten
>   Of the sun.
> —STANLEY WOOD.

TIME **forbade a** side excursion from Durango to the Mancos Cañon, though we were extremely anxious to make it,—*I* because I had been there before, and the rest because they were eager to see what I had told them of.

The Rio Mancos is the next tributary of the Rio San Juan west of the Rio de la Plata. When, in 1874, I was a member of the photographic division of the United States Geological and Geographical Survey, one of the main objects of our trip was the exploration of this remote corner of the State, where we had vaguely heard of marvelous relics of a bygone civilization unequaled by anything short of the splendid ruins of Central America and the land of the Incas. After traversing the frightfully rugged trails of the San Juan and La Plata mountains, therefore, a portion of our party came out on the southern margin of the mountains, and, despite the smoldering hostility of the Indians, with which the region was filled, headed southward into the long deserted cañons. There were five of us, altogether,—Mr. W. H. Jackson (from whose skillful camera came many of the illustrations that grace my present text), the famous Captain John Moss, who went with us as "guide, philosopher and friend," myself and two mule-packers.

The trail led from **Parrott City, then a nameless prospect** camp, washing gold without a thought of the silver ledges to be developed later there, over to Merritt's pleasant ranch on the upper Rio Mancos, then across rolling grass land and through groves of magnificent lumber pines, a distance of about fifteen miles. Spending one night at the ranch, sunrise the next morning found us eager to enter the portals of the cañon and the precincts of the area within which glorious discoveries in anthropology allured our imagination and made light the toil and privation of the undertaking.

ANIMAS CAÑON AND THE NEEDLES.

Not five hundred yards below the ranch we came upon our first find,—mounds of earth which had accumulated over fallen houses, and about which were strewn an abundance of fragments of pottery, variously painted in colors, often glazed within, and impressed in various designs. Later the perpendicular, buttress-like walls that hemmed in the valley began to contract, and that night we camped under some forlorn cedars, just beneath a bluff a thousand feet or so in height, which, for its upper half, was absolutely vertical. This was the edge of the green table-land, or *mesa verde*, which stretches over hundreds of square miles, and is cleft by these cracks or cañons, through which the drainage of the northern uplands finds its way into the Rio San Juan.

In wandering about after supper, something like a house was discerned away up on the face of this bluff, and two of us clambered over the talus of loose débris, across a great stratum of pure coal, and, by dint of much pushing and pulling, up to the ledge upon which it stood. We came down satisfied, and next morning Mr. Jackson carried up our photographic kit and got some superb negatives. There, seven hundred measured feet above the valley, perched on a little ledge only just large enough to hold it, was a two story house made of finely cut sandstone, each block about fourteen by six inches, accurately fitted and set in mortar now harder than the stone itself. The floor was the ledge upon which it rested, and the roof the overhanging rock. There were three rooms upon the ground floor, each one six by nine feet, with partition walls of faced stone. Between the stories was originally a wooden floor, traces of which still remained, as did also the cedar sticks set in the wall over the windows and door; but this was over the front room only, the height of the rocky roof behind not being sufficient to allow an attic there. Each of the stories was six feet in height, and all the rooms, upstairs and down, were nicely plastered and painted what now looks a dull brick red color, with a white band along the floor like a base-board. There was a low doorway from the ledge into the lower story, and another above, showing that the upper chamber was entered from without. The windows were square apertures, with no indication of any glazing or shutters. They commanded a view of the whole valley for many miles. Near the house several convenient little niches in the rock were built into better shape, as though they had been used as cupboards or caches; and behind it a semi-circular wall inclosing the angle of the house and cliff formed a water reservoir holding two and a half hogsheads. The water was taken out of this from a window of the upper room. In front of the house, which was the left side to one facing the bluff, an esplanade had been built to widen the narrow ledge and probably furnish a commodious place for a kitchen. The abutments which supported it were founded upon a smooth, steeply-inclined face of rock; yet so consummate was their skill in masonry that these abutments still stand, although it would seem that a pound's weight might slide them off.

Searching further in this vicinity, we found remains of many houses on the same ledge, and some perfect ones above it quite inaccessible. The rocks also bore some inscriptions. Many edifices in the cliffs escaped our notice. The glare over everything, and the fact that the buildings, being formed of the rock on which they rested, were identical in color with it, increasing the difficulty made sufficiently great by their altitude.

Leaving here, we soon came upon traces of houses in the bottom of the valley, in the greatest profusion, nearly all of which were entirely destroyed, and broken pottery everywhere abounded. The majority of the buildings were square, but many round, and one sort of ruin always showed two square buildings with very deep cellars under them and a round tower between them, seemingly for watch and defense. In several cases a large part of this tower was still standing. The best example of this consisted of two perfectly circular walls of cut stone, one within the other. The diameter of the inner circle was twenty-two feet and of the outer thirty-three feet. The walls were thick and were perforated apparently by three equi-distant doorways. At that time we concluded this double-walled tower (later triple-walled structures of the same sort were met with) must have had a religious use; but since then I have wondered whether all of these round buildings above ground (save some which manifestly were watch towers) were not used as storehouses for snow. It was a country of long droughts and hot summers. The double or triple walls, with spaces of dead air between would make excellent refrigerators.

These groups of destroyed edifices, occupying the bottom-land, were met with all day; but no other perfect cliff-houses were found until next morning, when a little cave high up from the ground was found, which had been utilized as a homestead by being built full of low houses communicating with one another, some of which were intact, and had been appropriated by wild animals. About these dwellings were more hieroglyphics scratched on the wall, and plenty of pottery, but no implements. Further on were similar, but rather ruder, structures on a rocky bluff, but so strongly were they put together that the tooth of time had found them hard gnawing; and, in one instance, while that portion of the cliff upon which a certain house rested had cracked off and fallen away some distance without rolling, the house itself had remained solid and upright. Traces of the trails to many of these dwellings, and the steps cut in the rock, were still visible, and were useful indications of the proximity of buildings otherwise unnoticed. Yet, despite our watchfulness, Mr. Holmes' party, which went next year to study the details of the broad prehistoric picture our rapid trip sketched out, brought to light several fine buildings, high above the valley, in some of which valuable implements and utensils were discovered. None of them were so high, though, or in better condition than one of our prizes this second day.

Keeping close under the mesa, on the western side (you never find houses on the eastern cliff of a cañon, where the morning sun could not strike them full with its first beams) one of us espied what he thought to be a house on the face of a particularly high and smooth portion of the precipice, which there jutted out into a promontory, up one side of which it seemed we could climb to the top of the mesa above the house, whence it might be possible to crawl down to it. Fired with the hope of getting some valuable relics of household furniture in such a place, one of the gentlemen volunteered to make the attempt, and succeeded. He found it well preserved, almost semi-circular in shape, of the finest workmanship yet seen, all the stones being cut true, a foot wide, sixteen inches long and three inches thick, ground perfectly smooth on the inside so as to require no plastering. It was about six by twenty feet in interior dimensions and six feet high. The door and window were bounded by lintels, sills and caps of single flat stones. Yet all this was done, so far as we can learn, with no other tools than those made of stone, and in such a place that you might drop a pebble out of the window 500 feet plumb.

Photographs and sketches completed, we pushed on, rode twenty miles or more, and camped two miles beyond Unagua springs. There were about these springs, which are at the base of the Ute mountain, the tallest summit of the Sierra ù Late, formerly many large buildings, the relics of which are very impressive. One of them is two hundred feet square, with a wall twenty feet thick, and inclosed in the center a circular building one hundred feet in circumference. Another, near by, was one hundred feet square, with equally thick walls, and was divided north and south by a very heavy partition. This building communicated with the great stone reservoir about the springs. These heavy walls were constructed of outer strong walls of cut sandstone, regularly laid in mortar, filled in with firmly packed fragments of stone. Some portions of the wall still stand twenty or thirty feet in height, but, judging from the amount of material thrown down, the building must originally have been a very lofty one. About these large edifices were traces of smaller ones, covering half a square mile, and out in the plain another small village indicated by a collection of knolls. Scarcely anything now but white sage grows thereabouts, but there is reason to believe that in those old times it was under careful cultivation. Evidently these thick walls were the foundations of old terraced pueblos, an unusually large community having grown up about these plentiful springs, just as at Taos, San Juan, Zuñi, and the present Moqui villages in Arizona.

Our next day's march was westerly, leaving the mesa bluffs on our right and gradually behind. The road was an interesting one, intellectually, but not at all so physically — dry, hot, dusty, long and wearisome. We passed a number of quite perfect houses, perched high up on rocky bluffs, and many other remains. One occupied the whole apex of a great conical bowlder, that ages ago had become detached from its mother mountain and rolled out into the valley. Another, worth mention, was a round tower, beautifully laid up, which surmounted an immense bowlder that had somehow rolled to the very verge of a lofty cliff overlooking the whole valley. This was a watch-tower, and we learned afterward that almost all the high points were occupied by such sentinel boxes. From it a deeply worn, devious trail led up over the edge of the mesa, by following which we should, no doubt, have found a whole town. But this was only a reconnoissance, and we could not now stop to follow out all indications.

Not far away the odd appearance of a cliff attracted my attention, and leaving the party I rode over the bare, white, rocky floors which capped all the low, broad ridges, to find a long series of shallow grottos in the escarpment filled with houses, some of which were roofed over, but most consisting simply of walls carried to the ceiling of the light, dry cavern in the sandstone, often only one or two houses occupying each of the small caves, whose openings were in the same water worn stratum, and only a few feet or yards apart. Still more curious examples of these cave-dwellings have been seen since in the same neighborhood, and lower down. For example, on the San Juan, in 1875, Holmes and Jackson discovered, half way between top and bottom of a bluff where a stratum of shaly sandstone had been weathered and dug out to a depth of six feet, leaving a firm floor and a projecting ledge overhead, a continuous row of buildings, though none have their front walls now remaining. Doorways through each of the dividing walls afforded access along the whole line. A few rods up stream a little, niched cave-house, 14x5x6, divided into two equal compartments; a small, square window, just large enough for one to crawl through, was placed midway in the wall of each half. "We well might ask whether these little 'cubby-holes' had ever been used as residences, or, whether, as seems at first most likely, they might not have been 'caches,' or merely temporary places of refuge. While, no doubt, many of them were such, yet in the majority the evidences of use and the presence of long-continued fires, indicated by their smoke-blackened interiors, prove them to have been quite constantly occupied. Among all dwellers in mud-plastered houses, it is the practice to freshen up their habitations by repeated applications of clay, moistened to the proper consistency, and spread with the hands, the thickness of the coating depending upon its consistency. Every such application makes a building perfectly new, and many of the best sheltered cave-houses have just this appearance, as though they were but just vacated."

The grandest of all these cave shelters, perhaps, was that in the

7*

Montezuma cañon, the main building of which was forty-eight feet long, and built of well smoothed stones. "In the rubbish of the large house," says the report, "some small stone implements, rough, indented pottery in fragments, and a few arrow-points were found. . . . The whole appearance of the place and its surroundings indicates that the family or the little community who inhabited it were in good circumstances and the lords of the surrounding country. Looking out from one of their houses, with a great dome of solid rock overhead, that

SILVERTON AND SULTAN MOUNTAIN.

echoed and re-echoed every word uttered with marvelous distinctness, and below them a steep descent of one hundred feet, to the broad, fertile valley of the Rio San Juan, covered with waving fields of maize and scattered groves of majestic cottonwoods, these old people, whom even the imagination can hardly clothe with reality, must have felt a sense of security that even the incursions of their barbarous foes could hardly have disturbed."

But I cannot linger over these extremely interesting and instructive ruins, nor stop to tell of the variety and skill shown in their architecture, in their storage of water and food, in their means of defense, in their manufacture of utensils, and the art with which their life was adorned. Out of the hundreds of leveled pueblos, cave-houses, towers, water reservoirs and wasted fields which once bore bountiful harvests, I have only culled one here and there. I may say that not only every cañon which cuts down through the mesa to the Rio San Juan and into all of its lower tributary valleys, but many of the plateaus between, are occupied by the ruins which show an Indian occupation previous to the present savages, and of a different rank, if not of another race.

Particularly accessible to the ordinary tourist are the ruins to be seen in the Animas valley, about twenty-five miles south of Durango. These are said to consist of a pueblo three hundred and sixteen feet long by nearly one hundred wide, which evidently rose to the height of many stories. Some of the lower rooms in this great house are still standing, and skeletons and relics of great interest have been taken from them. In the center of the ruins is a subterranean, cistern-like chamber, described as about sixty feet in diameter, and plastered everywhere within with hard cement. This, probably, was the main *estufa* of the village. Other lesser ruins and remains of farming operations are scattered about the vicinity, and are well worthy of exploration.

Just who and what were these aborigines (if so they were, which is very doubtful), opinions differ; but that in the Village Indians of New Mexico and Arizona we see to-day their lineal descendants, seems indisputable.

Traditions are few, that have any value, but the partial and imperfect researches which have already been made in the southwest enable us to make out dimly some strangely tragical scheme of history for this race of men whose sun set so long ago.

It is evident, for example, that the most ancient of these prehistoric ruins are those found along the immediate banks of the water-courses in the valleys. There the forerunners of the troublous times to come dwelt in peace and prosperity among their fields, which seem to have stretched over many times the area of land now possible to be cultivated. There is no question, indeed, that in those days rains were more frequent and the climate far more favorable to agriculture than at present. But how many generations—how many centuries—ago was this? And how did the change of climate, which turned the fertility of the land into desolation, come about—by slow degrees, through sudden cataclysm, or with comparatively rapid advance? Probably gradually.

But it does not seem to have been as the result of meteorological disfavor that they abandoned their populous pueblos in the pleasant valleys and began to build refuge homes in the niches of the cañon's wall, or on the crest of inaccessible mesas. From the mountainous north came enemies they were unable to resist, and which devastated their fields and laid waste their towns, as we have seen at Ojo Caliente, and as is written in the ruins of a hundred spring-side pueblos throughout the San Juan valley. No doubt they still cultivated their fields as well as they could between the times of attack, building temporary summer-houses and spending the idle winter in their rocky fastnesses, or retreating to them when warned of an attack. Their watch-towers on every exposed point, tell how sharp and incessant was the lookout they kept against the well-mounted and savage nomadic tribes, the prehistoric Utes and Apaches and Navajos, who were to them as the Scythians and the Vandals and Goths to the weakened empire of effeminate Rome.

But after a time a breathing space seems to have come to the harassed people, and they felt themselves safe to return to their ancient valleys and reinhabit and recultivate them. Certain houses, built upon the substratum of older fallen structures, seem to show this new era of reoccupation, which in some places lasted only a short time before enemies and drought together compelled complete abandonment, while in other more southern strongholds were founded the pueblos that still exist, at Taos, Acoma, Zuñi, and on the Moqui mesas.

When, some day, you can ride down the Mancos in a railway car and get flying glimpses of the ruined houses—if your eyes are sharp to see and your mind quick to apprehend,—do not forget how populous was this dry and garish valley during those bygone days, when the Crusaders were waking up Europe, and all that was known of America was that the Basque fishermen went to the fog-banks of an icy western coast to catch codfish. I am more sure of your interest here, though, than in many other far-paraded precincts of this marvelous realm, I am taking you so swiftly through in my pilgrimage on wheels. And I cannot enforce my point better,—leave an impression more lasting and graceful on your minds of those gentle shepherds and husbandmen (but no less brave warriors), who were here so long before us, than by giving you the poem my clever-brained and genial friend has written in Swinburnian measure about them

> " In the sad South-west, in the mystical Sunland,
>   Far from the toil, and the turmoil of gain;
> Hid in the heart of the only—the one land
>   Beloved of the Sun, and bereft of the rain;
> The one weird land where the wild winds blowing,
>   Sweep with a wail o'er the plains of the dead,
> A ruin, ancient beyond all knowing,
>   Rears its head.
>
> " On the cañon's side, in the ample hollow,
>   That the keen winds carved in ages past,
> The Castle walls, like the nest of a swallow
>   Have clung and have crumbled to this at last.
> The ages since man's foot has rested
>   Within these walls, no man may know;
> For here the fierce grey eagle nested
>   Long ago.
>
> " Above those walls the crags lean over,
>   Below, they dip to the river's bed;
> Between, fierce wingéd creatures hover,
>   Beyond, the plain's wild waste is spread.

No foot has climbed the pathway dizzy,
   That crawls away from the blasted heath,
Since last it felt the ever busy
   Foot of Death.

" In that haunted Castle—it must be haunted,
   For men have lived here, and men have died,
And maidens loved, and lovers daunted,
   Have hoped and feared, have laughed and sighed—
In that haunted Castle the dust has drifted,
   But the eagles only may hope to see
" What shattered Shrines and what Altars rifted,
   There may be.

' The white, bright rays of the sunbeam sought it,
   The cold, clear light of the moon fell here,
The west wind sighed, and the south wind brought it,
   Songs of Summer year after year,
Runes of Summer, but mute and runeless,
   The Castle stood; no voice was heard,
Save the harsh, discordant, wild and tuneless
   Cry of bird.

" The spring rains poured, and the torrent rifted
   A deeper way;—the foam-flakes fell,
Held for a moment poised and lifted,
   Down to a fiercer whirlpool's hell.
On the Castle tower no guard, in wonder,
   Paused in his marching to and fro,
For on the turret the mighty thunder
   Found no foe.

" No voice of Spring,—no Summer glories
   May wake the warders from their sleep,
Their graves are made by the sad Dolores,
   And the barren headlands of Hoven-weep.
Their graves are nameless—Their race forgotten,
   Their deeds, their words, their fate, are one
With the mist, long ages past begotten,
   Of the Sun.

" Those castled cliffs they made their dwelling,
   They lived and loved, they fought and fell,
No faint, far voice comes to us telling
   More than those crumbling walls can tell.
They lived their life, their fate fulfilling,
   Then drew their last faint, faltering breath,
Their hearts, congealed, clutched by the chilling
   Hand of Death.

" Dismantled towers, and turrets broken,
   Like grim and war-worn braves who keep
A silent guard, with grief unspoken,
   Watch o'er the graves by the Hoven-weep.
The nameless graves of a race forgotten;
   Whose deeds, whose words, whose fate are one
With the mist, long ages past begotten,
   Of the Sun."

XVI

ON THE UPPER RIO GRANDE.

> O wonderful, wonderful, and most wonderful wonderful, and yet again wonderful and after that out of all whooping.
> —Merchant of Venice III, 2.

OFF to Del Norte and Wagon Wheel Gap! That meant a long run. We might have gone afoot across the Cunningham Pass and down the Alpine fastness of the Rio Grande's birthplace almost as speedily as the train would take us, back to Durango, over the heights and glories of Toltec, down the mazy labyrinth of the Whiplash, and across the sheep pastures of San Luis. But we were in no hurry, and by preparing had the jolliest time you can imagine the whole way. At Alamosa we bid a reluctant farewell to our three companions, the Artist, the Photographer and the Musician, who can no longer spare to us their society. But our prospective loneliness is mitigated by a new comer,—an old college friend. I shall introduce him to the reader as *Chum*, because that was the ordinary way in which we dispensed with his name.

> " A merrier man
> Within the limit of becoming mirth
> I never spent an hour's talk withal."

Here, the good-bye and the welcome given in the same breath, we change our cars to a new train headed westward toward the upper course of the Rio Grande, with its farms and mines and medicinal springs.

The track is laid right across San Luis park, which is to become, through irrigation, one of the richest agricultural regions in the world. By and bye, the dull line under the horizon began to form itself into trees, and among these we could distinguish the scattered log and adobe dwellings and the half-cultivated little farms of Mexican ranchmen. The bottoms of the Rio Grande now spread wide around us, with bushes and trees, and tall, rich grass, and a few miles further on we came to the town amid a group of picturesquely broken volcanic bluffs of great size. This is a sort of postern-gate for the San Juan mining region, and also for Lake City, Ouray and the San Miguel.

Only a postern-gate, for now the railway southward carries the passenger to Durango and Silverton, and the Salt Lake line makes an easy entrance to the northern slope of the Sierra Madre; but, a few years ago Del Norte was the last outfitting point for those going into all that region, and the first real civilization encountered on the return. Under the "boom" of this patronage the old Mexican ranch center became an American town of some size and importance almost ten years ago, and its people thought they were soon to be the metropolis of the southwest. But such has not yet appeared to be their destiny, and a snug, stirring little village of twelve or fifteen hundred people is all that the settlement has developed into. It is charmingly placed, and there is so much land along the river, both above and below, which is cultivated by both Mexicans and Americans (chiefly in the line of hay), and so many sheep, cattle and horses are owned and sold there, that this interest alone will support the village and enable it to grow slowly.

But pleasant Del Norte has more than this to rely upon. Twenty-eight miles back in the mountains of the Continental divide are the famous Summit gold mines. The richness of these mines (as they appear at present) is almost inconceivable—it equals the fabled *El Dorado* so many brave fellows have died in their effort to find. The railway express company, in the three months following the advent of the road at Del Norte, forwarded to the Denver mint $300,000 in gold bars. I have seen and handled many pieces of this reddish, rusty, honey-combed quartz, in which you could see the gold as thickly and plainly as the pepper on sliced cucumbers. There were streaks of it, maybe half an inch wide, where the material was more than half its weight, pure, visible gold.

Prospecting on South mountain in 1874 (or before that) men found these ledges, and various claims were staked off, and, in 1875, stamp mills were erected, which at once began grinding out thousands of dollars a day and saving only about sixty per cent. of the gold, the remainder running off in the tailings because it was too coarse and heavy to be caught quickly by the mercurial batteries; this was enough to set fire to the tinder of the gold-seeking population, which is always ready to stampede to a new camp, and in 1876 a great rush to the Summit district happened. The whole region was quickly put under claim-stakes and a dozen respectable mining beginnings were made. Among these was a group of claims, more or less worked, which became the property of a corporation called the San Juan Consolidated Mining Company. Their principal mine was the "Ida," and their most intelligent stockholder was Judge, now United States Senator, Thomas Bowen. He came to this region from Arkansas an exceedingly poor man, though in early life he had been a wealthy planter. Elected a justice of his judicial district, he plodded on foot from county to county, too poor to own a horse. For seven long years, the story goes, he put all his money into prospecting, and at last turned up here at the Summit. Watching the way in which the "Consolidated" property was being handled, he concluded that its managers were not on the right track and would speedily come to a halt; furthermore, he had faith that he could right the mistake if he had the power.

As he anticipated, the stock of that company went down to nothing. No further back than the winter of 1880-81, its shares were played at poker in Del Norte, and passed over the bars of saloons at the rate of two drinks for one share. Bowen quietly gathered them in, getting $300,000 worth, it is stated, for $75, or one-fourth of one mill on the dollar. Two or three others saved up smaller amounts. When the Judge had secured a controlling interest, he set on foot a scheme of new development, and very shortly struck this fabulously rich vein. He persuaded friends in Denver to erect a mill on terms which have resulted in the biggest profits a stamp-mill ever paid its manufacturer, I fancy, and Bowen suddenly found himself a Crœsus. He

CLIFF DWELLINGS.

had been heavily in debt, and some of the scores against him had long been charged to loss by his creditors, but he paid them all without noticing the drain upon his uncounted coffers. Having fought the demon of poverty in its most tenacious forms, for so many years, this sudden affluence did not spoil him, but he glories in it like a boy, and is never more pleased than when he can make it tell for the surprise and happiness of some old companion still in the grip of misfortune.

But "Bowen's bonanza" is not the only one. There are others of perhaps equal merit close by, and I have no doubt many more will be discovered. For, in spite of all the bullion which has this year been produced, these mines are as yet in their infancy. I suppose the measure of half a mile would include the total length of underground workings in all of them together. Who shall say what the future may not disclose?

Half a dozen miles across the mountain from the Summit is a flourishing little settlement of prospectors who believe they have struck a profitable lode of silver-galena; and still farther beyond, among the springs of the Rio San Juan, lie the Cornwall silver mines, where much work has been done. The principal properties are on the Perry lode, which gives sulphuret of silver. Other ores there vary from this, however, and are said to be best suited to the lixiviation process. A smelter has been purchased for that locality. Judge Jones, so well known all over southern Colorado for his steady allegiance to everything which savors of "San Juan" and for his equal hatred of whisky, has large visions of future wealth out of this district.

Now the whole of these mines and trials for a mine are so much grist to Del Norte's mill. So long as they keep men digging, so long she will thrive exceptionally and remain an important feeder to our railway.

The scenery along the Rio Grande, above Del Norte, is very fine, and has always the zest of human interest in the quaint ranches of the Mexican farmers, whose women and children flock out to see every train go by. Terraced steeps bound the river-valley where the farms are, at a little distance, on the right, while rounded pine-clad hills slope upward on the left. We see this part of the river on our way to Wagon Wheel Gap, a place for which, if I were writing a separate chapter, I should adopt as a motto the words of Exodus: "And they came to Elim, where were twelve wells of water and three score and ten palm trees, and they encamped there by the waters."

"But what is Wagon Wheel Gap, and how did it get such a name?" asks the Madame.

"The gap," it is thereupon explained, "is a noble gateway, thirty miles west of Del Norte, through which the Rio Grande breaks out of the confinement of its youth in the San Juan mountains; and I heard only yesterday how it come by its name, from the great and good Judge Jones, whose narratives most happily combine both facts and fancies.

"You will remember what we heard of the band of men who went into the San Juan mines four-and-twenty years ago, under Colonel Baker. Well, there was a part of that story you have not heard yet. It seems that the party was composed of Northern and of Southern men in nearly equal numbers. When they heard that war had broken out between the Northern States and the hoped for 'Confederacy,' there was added to the woe of disappointment, diminished food, and the fear of Indians, the bitterness of a little civil war among those who previously had been compatriots and friends. It was a miserable little copy of the great struggle, but it resulted in disproportionate sorrows, for a panic ensued, in which the men of the party broke up and scattered out of the mountains by every available passage, a prey to double the dangers which would have menaced them had they stayed together. Some tried to take their wagons out piecemeal over Cunningham Pass. Putting them together on the eastern side, they worked their way down to Del Norte, Fort Garland, and so to Santa Fe, or around to Denver. But they often broke down, and a relic of this panic-stricken flight, in the shape of a large wagon wheel, found by Judge Jones, served to give the place its peculiar name. To distinguish it from other gaps in the range it was spoken of as 'the gap where the wagon wheel was found,' which soon, by natural process of curtailment, condensation and transposition, became 'Wagon Wheel Gap,' and Wagon Wheel Gap it is, even unto this day."

The gap itself is a cleft through a great hill; or it is two half hills (for they stand not squarely opposite one another, but with somewhat overlapping ends only) each vertically faced and uprightly seamed on the river side, but sloping away into a grassy ridge behind. The southern end of the bluffs, which also is the farthest up stream, is narrow and tower-like, but the other, rounding out and swelling high, in the center has a breadth of half a mile or more, the river washing its bowed base. Of about the same height as the Palisades of the Hudson, and like them marked with vertical lines of cleavage, this bluff of reddish volcanic rock would bear a striking resemblance to that great monument of the Plutonic reign on the Atlantic slope, did its façade present a straight front; but in this swelling front is where it exceeds its eastern rival, for one gets added pleasure from the perspective of the massive battlement retreating right and left in grand curvature.

The gap is wide enough not to pen the water into a very narrow flood, so that only a slight exaggeration of the always lively current occurs. This is enough, however, to make these ripples a favorite spot for the splendid trout in which the whole upper part of the river abounds, and I suppose four or five thousand pounds of these gamy fish are taken every year.

Right at the foot of the talus stands a group of connected log-cabins forming a comfortable hotel. As we are bound for the hot springs, we do not remain here, but climbing into the spring wagon in waiting ride southward for a mile back into the hills to a second little *cuchara* of a valley where are the springs themselves and the hotel, bath houses and accessories belonging to the sanitarium they have created. This hotel is delightfully home-like in its excellence of bed and board.

Persons go to Wagon Wheel Gap seeking recreation and for recovery from ill health. If the first is their object (as it has been in the case of the late Secretary of the Treasury,—one of the hardest working men for ten months in the year, in the United States) they have a pleasant home and good company and no end of out-door fun from which to choose.

There are little hills near by which they may begin on, and taller bluffs beyond, where they may make perfect their practice; while far away stand the "ultimate heights" of the Sierra Madre,—unbroken masses of snow when we last beheld their spear-pointed peaks. There are ponies they may ride, and donkeys for the children. There are single buggies and phætons for the ladies and carryalls for the whole family. There are geologizing, botanizing and general natural history to invite study in endless variety.

Then there is good sport. That noblest of the deer race,—the elk— still haunts the upland pastures and mountain glades. The black-tailed deer is to be found lurking in the aspens, and, if you are a good climber, you may enjoy the very next thing to Alpine chamois shooting in the arduous chase of the mountain sheep. As for fishing, there is no stint to it in the proper season. I know no place in Colorado where the fly-fisher will have better sport and the angler, though uninstructed in the wiles of Walton, get better results.

But it is to invalids that the hot springs especially appeal, holding out all these pleasures for their delectation as gradually they regain their health sufficiently to practice and enjoy them. Long ago these beneficent waters were resorted to by the Indians for healing. A trail, not yet obliterated, ran across the hills to Pagosa Springs, which were called The Big Medicine, while these waters were known as The Little Medicine.

Though of less volume than single springs at Pagosa, the springs at Wagon Wheel Gap pour out nearly as much water, since there are thirty or more of them in the sheltered basin which makes a natural sanitarium. Some are icy cold, others tepid, others extremely hot. They are diverse also, in respect to their mineral constituents, nearly every known variety of *spa* water being represented more or less closely. Only a few of the springs are utilized, however, those being selected which seem to have the most powerful curative properties. These are principally three,—and are known as Nos. One, Two and Three. The analysis of them published, though imperfect, but serves to show the general character of each, and reads as follows, the proportions being thousandths of a given bulk of the water:

	No. 1.	No. 2.	No. 3.
Sodium Carbonate	69.42	Trace.	144.50
Lithium Carbonate	Trace.	Trace.	Trace.
Calcium Carbonate	13.08	31.00	22.42
Magnesium Carbonate	10.91	5.10	22.42
Potassium Sulphate	Trace.	Trace.	Trace.
Sodium Sulphate	23.73	10.50	13.76
Sodium Chloride	29.25	11.72	23.34
Silicic Acid	5.73	1.07	4.75
Organic Matter	Trace.	Trace.	....
Sulphureted Hydrogen	Trace.	12.00	....
Total	152.12	71.39	218.77

The largest of these is the "Number One," and from it is drawn the water for the plunge-bath and to the private bath-rooms, which is used in the largest number of cases where the disease affects the nerves, blood or skin. An oval basin twenty feet or more in length has been excavated and tastefully walled up. Here the spring bubbles up most copiously, at a temperature of 150 degrees Fahrenheit, and sends off an incessant cloud of steam if the air is at all cold. From this tank the water is conducted in an open trough to the bath-houses, losing about 30 degrees of its heat on the way. This introduces it into the baths at the quite endurable temperature of 120 degrees, but fills the compartment with a cloud of vapor, so that the patient breathes in the chemically-laden moisture with every inhalation. Besides this, while in the bath, the fresh hot water is drunk in big draughts. The invalid is thus soaked out and in with the healing fluid; the pores of his skin, all the passages of his head and chest, his stomach and secretory organs feel the touch of the water and eagerly absorb the medicinal elements it contains.

Astonishing results have come from a steady continuance of daily baths and sweatings. They will tell you instances—and vouch for them, too, by incontestable testimony—of men brought there utterly helpless and full of agony from inflammatory rheumatism or neuralgia, who, in a week, were able to walk about and help themselves, in a fortnight were strolling about the valley erect and comfortable, in a month went to work. Three months of faithful self-treatment, it is confidently promised, will set straight the most chronic and painful cases of such invalidism. Many a miner now digging in the wintry mountains passed from almost certain death to exuberant strength through this Siloam, and evidences are being multiplied of the startling efficacy of these springs with every additional season.

Then there is the dreadful list of cutaneous diseases and disorders of the blood, headed by the fiendish heritage of syphilis. To such cases, because their disease is communicable, are set apart separate baths. Here, again, utter helplessness and the awful suffering which tempts to suicide, are relieved by a few weeks of steady application; and not merely relieved, as the druggist's medicines might in some cases do, but, it is claimed, thoroughly and healthfully cured. But this last claim, in the case of pronounced syphilis, needs the confirmation of longer trial. I am willing to admit that the two or three, or four years which have elapsed since certain persons have gone away restored to health, have shown no recurrence of the symptoms; but I should like to know that the same thing could be said indisputably half a century hence before I would be willing to admit as proven that this spring or any other could lay low forever the head of a malady which it has hitherto baffled medical science to eradicate wholly from an infected system. But even if this final, full glory shall never be attained by the Wagon Wheel

Springs, it is enough joy for the present that they are able to alleviate its miseries and even temporarily check the havoc of body and soul. I have said so much upon this point, because, it being impossible to deny the existence of the evil, it is every man's duty to aid in spreading a knowledge of any method of relief or cure.

It is in the two classes of diseases above mentioned that the little cold spring comes into play as a useful beverage. Dyspeptics, however, bathing with less assiduity than their more unfortunate brethren, affect the hot soda water (Number Three), which bubbles up in a strong, hot fountain at the head of the gulch, and surrounded by chalybeate springs of various qualities. This is the water, too, which the mildly ailing like to sip, perhaps mingling a little lemon juice and sugar with it to make a foaming compound grateful both to the taste and the system,—a union rare in these days of doctors.

So, thanking God, we were in no need of the Little Medicine for health, but could enjoy its delicious warmth and fragrance as pleasure unalloyed; but profoundly grateful that for humanity in worse luck there was such an Elim in the desert of our degeneracy, we bade adieu to pleasant, sunny, warm-hearted Wagon Wheel and its jolly landlord, —Mr. McClelland, our compliments, and your good health, sir, in something stronger than mineral water!

Down at the station we went fishing, partly for fun, partly with the urgency that set the boy digging out the woodchuck. Marvelous stories —regular fish stories of eight-pound trout caught on a seven-ounce rod,— had been dinned into our ears; and as for me, I half believed them (for I remembered the splendid fellows we used to snatch from the White Water at pretty Irene cañon, up above Antelope park). Our ambition was not to repeat such performances, but to get one or two of, say a pound and a half each; while the Madame said she'd be thankful if we had a few little ones not worth weighing, by dinner time. So Chum and I went down stream rod in hand.

Having floundered round on the slipping bowlders for awhile without sitting down, we struck a couple of good-sized pools at the head of a riffle; Chum took the upper, I the lower. Making my way out near to mid-stream, I took up my station behind a large flat rock that stood about a foot out of water, and busied myself sending a "coachman" and a "professor" out into my domain with a little hope that I might induce something out of the inviting pool. Before I had been there five minutes, a yell from Chum caused me to look his way. His Bethabard was beautifully arched, and at the end of twenty feet of line something was helping itself to silk.

"I've got him; he's a whopper."

"That's the pound and a half I promised you," I answered, as a beautiful fellow shot across stream not three yards above me. "But you'll lose him in that current."

"I know it, unless I work him down your way."

"Come on with him—don't mind me." I reeled in, climbed on the rock, and sat down to see the fun. The noble fellow made a gallant fight, but the hook was in his upper jaw, and it was only a matter of time when he would turn upon his side. Working him down stream, through my pool and round into the quieter water near shore, was the work of ten minutes at least, the captive, seeming to readily understand that still water was not his best hold, kept making rushes for the swift current; but each time he was brought back, and soon began to weaken under the spring of the lithe toy in Chum's hand. Fifteen minutes were exhausted when the scale hook was run under his gills and he registered one pound twelve ounces.

Apologizing for creating a row in my quarters, Chum went back to his old place, while again I tried my luck. About five minutes elapsed when I heard another not to be mistaken yell.

"I've got another—he's bigger than the first."

"Yes, I see you have—I think it's infernally mean."

"I know it is, but I can't help it. I've got to come down there again."

"Well, come on," and I sat down again to watch the issue. The struggle was not so brave, though the fish, when brought to scale, weighed half a pound more than the first. While we were commenting on this streak of luck, we noticed a change in the water, its partially clear hue began to grow milky, and in less time than it takes to tell it, a bowlder six inches under the surface was out of sight.

"We might as well go to dinner, no trout will try to rise in that mud," and I reeled up with the reflection that the next best thing to catching a trout is to see one captured by one who knows how.

The next day we had another try. Chum crossed the river, and then we slowly walked down to a magnificent pool a mile below. Here were a party of half a dozen gentlemen, to one of whom I called out just as we came within hearing.

"Have you got him?" The inquiry was made on the score of good fellowship; the bend of his split bamboo, the tension of his line, and the whirr of his reel indicated the first stage.

"I've hooked him, and he's no sardine, I tell you—whoa, boy, gently now," as a sudden rush strung off full twenty feet of line. "Whoa, boy, be easy now; gently, now; come here; whoa! confound your picture! whoa, boy, gently, so, boy."

"May be you think you are driving a mule," came from one of the anglers.

"Oh, no! I'm trying to lead one—whoa, boy, whoa, boy, gently, now, none of your capers—whoa! I tell you!" as a renewed and vigorous dash for liberty threatened destruction to the slender tackle. "No you don't, old fellow, so, boy, that's a good fellow," and showing his back near the surface, the captive exhibited twenty inches (at a guess) of trout.

"By George, he's a beauty," came from behind us.

I had allowed my flies to float down stream and had backed out to give room for fair play. It was a long fight, but his troutship finally showed side up, and was gently drawn ashore, the water turned out of him, and he drew down the scale three pounds to a notch.

These are only "pointers" for the angling fraternity. As for our own luck that day—well, we had good trout left in the ice-box for a week.

It was our fortune throughout this trip to mix experiences by the most sudden transitions. It did not seem strange to us, therefore, that from the gold-fields and trouting and jollity of this beautiful valley, we should go at a jump into the coal districts east of the range.

XVII

EL MORO AND CAÑON CITY.

> For Knights no more in modern days bestride
> Their Rosinante and across the hills
> Ride, by my halidome, to succor maids
> Or couch a lance against an amorous foe;
> Instead, within a Pullman Palace Car
> At ease reclining and at peace with all,
> We conquer space while romance groweth dull
> Under the languor of the April air.
> —WILLIAM E. PABOR.

"EL MORO, sir,—breakfast nearly ready, sir!"

I had only closed my eyes an instant before, I was sure, yet then I had been lying quietly in the station at Alamosa, away over on the other side of the Sangre de Cristo. I couldn't remember anything of the transition. Then it was night; now it was morning. Time and space had been an utter blank for ten hours and a hundred miles.

Drawing aside my window curtain and gazing out over gray plains, my eyes caught instantly the bluish outlines of a grander castle and fortress than ever traveler on Rhine or Danube glanced upon. Almost a twelve-month before the Madame and I had spent a sunny day at St. Augustine, where the old square, four-bastioned fort stood grim on the shore of a foam-flecked and laughing sea. Here was a copy of that fortress a thousand times as large,—sloping walls, outer works, bastions, towers and all; you might almost see the huge guns, standing rigid and ready on the magnificent parapet. This was *El Moro*. It was miles and miles away, and a hundred cities had room to cluster about it and take its name without crowding one another. It is the most magnificent model of a half ruined, antique, but altogether glorious fortress in the whole wide world.

At El Moro are some of the great coal mines of this carboniferous state; but to put the matter Hibernically, they are half a dozen miles away up in the hills. We went up there on an engine which drew behind it a box-car load of Mexicans, men and women, who all squatted down on their heels on the car floor and were quite happy, chattering like magpies. It is a very miserable class of Mexicans, for the most part, that one sees in this part of the state. The fathers and mothers of most of them were *peons* of the wealthy Spaniards, who, under the old *regime*, owned all this region as pasturage for their flocks and herds.

The coal-bed is some hundreds of feet higher than the valley of the Purgatoire, where the village is; and it can be traced for thirty miles east and west, cropping out frequently and used all along by the farmers who live near it, as a supply of fuel. Throughout this extent it varies, of course, in thickness and quality, though in no great degree. Where the El Moro mines are opened, it runs from a vertical thickness of thirty feet to thirteen, nearly all of which is solid, merchantable coal. There are two thin streaks of what the miners term "bone"—something neither coal nor slate—and little patches of refuse, now and then, but no great amount.

I do not know how many hundreds of yards of underground tunnels we walked through, but it was an immense distance, yet the foreman said we had not been half way through it all. Everywhere the roof was high overhead and the walls solid coal, so that the men did not have to crawl on all fours or lie prostrate and dig, as I have seen done in some eastern mines, but could stand full height. All the product of the mine is taken by the pick, except a small amount cut by machines, which dig a horizontal trench, level with the floor, for five feet or so under the breast of the coal and across its whole breadth. This machine works with extreme rapidity, and after it is done, a couple of blasts will topple down as much coal as can be carted out in a day.

The output each month is nearly twenty-five thousand tons, about two hundred and seventy-five miners working. The piece-work method is adopted in paying, and though a smaller rate is paid per ton in mining than is usual in the Pennsylvania or Illinois mines, the earnings of the miners aggregate much larger than the average in the East. This is due to the superior ease with which this coal can be taken out, because of its softness and the roominess of the working chambers, rather than to anything better in the miners or greater diligence in working. One difficulty,

UP THE RIO GRANDE

indeed, arises from the ease with which high earnings can be made here, namely, that desirable workmen will labor only during the winter, or until they have made a "grub-stake," as they say, when they will go into the mountains prospecting for mines until their funds are exhausted, for, thus far, none of them have become suddenly wealthy.

At the mines, the Colorado Coal and Iron Company, which owns this property, have built a small village of *adobe* and wooden houses, in which the miners reside. About two hundred men were employed, many of whom had families. They were of almost every nationality, including some sixty Mexicans.

The El Moro coal is a true bituminous coal, producing a coke of excellent quality. It is asserted by its owners to be the best coal for making gas and for blacksmithing in the state; and it is used extensively for steam and metallurgical purposes. It is also said to be the only coal yet discovered in Colorado, lying east of the mountains, which can be profitably used for heating iron in furnaces, and for this it is equal to the best grades of eastern coal. About two-thirds of the product is made into coke.

The coke-works lie five miles from the mines and near El Moro, where there are three steam pumps, a fifty-horse power engine, crushing and washing machinery and two hundred and fifty coking ovens. A charge for each of these ovens, they told us, was about four and a half tons, and the yield of coke from each, after forty-eight hours of burning, is about two and a quarter tons, making a present total product of two hundred and eighty tons of coal a day.

The following shows the analysis of this coal and coke, and also that of the well known Connelsville coal and coke:

COAL.

	Water.	Vol. Matter.	Fix. Carbon.	Ash.	Sulphur.
El Moro	0.26	29.66	65.76	4.32	0.85
Connelsville	1.26	30.11	59.52	8.23	0.78

COKE.

	Fix. Carbon.	Ash.	Sulphur.
El Moro	87.47	10.68	0.85
Connelsville	87.26	11.99	0.75

This analysis of El Moro coal was made from selected specimens; the average of ash in the coke will probably run up to twelve to fourteen per cent.

When we were at Cucharas once before (I have omitted to mention it in its proper place), we ran over to Walsenburg, a neat little settlement in Huerfano Park, and a headquarters for a large sheep industry, and visited the coal mines of this company near by. They own a large tract of land there, containing three seams of coal, four, nine, and five feet in thickness. Only the thickest of these has as yet been developed, sending out about seventy-five thousand tons annually. This coal analyses as follows, "No. 1" being the four-feet seam, "No. 2" the nine-feet seam:

	No. 1.	No. 2.
Water	3.23	2.97
Vol. Matter	40.93	40.08
Fixed Carbon	49.54	48.67
Ash	6.30	8.28
	100.	100.
Sulphur	.62	.65

It was old traveled ground, for the major part of the distance between El Moro and Cañon City, on the Arkansas, forty miles above Pueblo; and as we were anxious to save all the time we could for the new regions in the west, where we were sure the most romantic experiences awaited, we decided for another night-run, disadvantageous as they were, compared with day journeys. The next morning after our visit to El Moro, therefore, found us at anchor in Cañon City.

"What is there to see about Cañon City?" Oh, quantities of things. Here is a list of what its *Record* keeps "set up" as its "advantages, natural and otherwise:"

"Soda springs, iron springs, hot soda baths, wide streets, excellent town site, immense water power, exhaustless coal fields, good water works, best building stone, splendid lime rock, iron mines, mica mines, lead mines, silver mines, oil wells, irrigating ditches, abundance of shade trees, peaches, plums, pears, apples, walnuts, grapes, vegetables, grain, flowers, bees, fifteen thousand dollar school house, twenty thousand dollar court house, Masonic temple, city government, low taxes, streets sprinkled, seven churches, theatre hall, first-class dentists, two newspapers, excellent physicians, good teachers, brick and stone stores, excellent society, protection from cold winds, immense stocks of goods, railroad communication, good ranches, stock ranges, excellent hotels, military college, and kindergarten."

Most of these items describe themselves, but others are worth mention. Right at the mouth of the Grand Cañon, which suggested the name, this site early attracted to a permanent home many of the earliest wanderers whom the famous Pike's Peak immigration of 1859 brought to the country. Half a century before them, though, Major Zebulon Pike had made a station for part of his troops on this spot, whence he reconnoitered the surrounding mountains.

Basing their calculations upon the fact that their settlement, which from the first was called Cañon City, was the last place to which the big emigration and freight wagons could come from the plains, the pioneers had large hopes of their town as the one entrepôt and supply-point for the mountains. Merchants came here and crammed great sheds with stocks of goods sold at wholesale, while forwarders were busy in organizing ox-trains to carry supplies into the mountains.

Then came the War of the Rebellion. All travel along the southern trail across the plains was cut off by the Indians, and immigration ceased, particularly from the Southern states, whence had come into this part of Colorado a large portion of the early settlers. More from lack of anything else to do than because of strong convictions on the

GRAPE CREEK CAÑON.

subject, some two hundred and fifty young men enlisted from here into the Union service and were sent to New Mexico on that campaign against Sibley, wherein Colorado's regiments distinguished themselves. While the war raged Colorado was at a standstill, and the settlers had hard shift to live, all goods having to come by the way of Denver, subject to great risk.

Then the war closed, emigration westward revived, and Cañon City, along with the rest of the region, took a new lease of life. A committee of Germans came from Chicago seeking a place for a colony of their compatriots, and were guided by Mr. Rudd until they hit upon the Wet Mountain Valley and located Walsenburg. A little later, hopes of a railway cheered the hearts of the southern Coloradoans. The Kansas Pacific Company sent engineers up the Arkansas to locate their road across the range. They surveyed to this point, estimated upon the cost of grading through the cañon, and over the range by two or three routes. Then they abandoned the locality and deflected to Denver. This was no sooner done than reports of the advance of the Atchison, Topeka and Santa Fe came to cheer the citizens, but disappointment ensued until early in the last decade the Denver and Rio Grande staked out their narrow gauge, and had the cars running regularly hither by the Spring of 1874.

All this time the town was slowly progressing and its vicinity being taken up for ranches, claimed for coal lands, quarried for building-stone and lime, and prospected for the precious metals. In 1879 came a "boom." Leadville flashed into sight and the Rosita and Silver Cliff district sprang to the front to rival it in excitement. To both these centers of rushing crowd Cañon City was one point of ingress. The town suddenly became thronged with men and women who were straining every nerve to get into the new regions, and with the undertow of a returning army of men disgusted with their reception and ill-luck, or jubilant over quick success, or going east only to return with machinery or goods or more money and friends.

To the task of catching toll from this careless, hurrying, fat-pocketed stream of humanity Cañon City set herself. The hotels charged big prices, lodging houses started up right and left, restaurants and boarding tents thrust out their signs every few steps, merchants urged your renewing your outfit at their replenished counters, and every form of wild amusement led to revelry and hideous headaches. Of all the wild towns it has been my fortune to see in the West, I think the Cañon City of those days was the worst. Ruffianism was the only fashionable thing, seemingly, and from the tangle-headed, dusty and drunken bull-whacker or the professional card-sharper to the staid old citizen, everybody had caught the spirit of a grand spree and the devil reigned.

But the railway was at last built through the cañon, the loud-swearing, quick-shooting teamsters followed their principals to the end of the track at Cleora, Salida, Buena Vista, and so on to Leadville itself; the gamblers and dance-houses and harlots followed; the host of railway laborers no longer made the village streets scenes of debauchery, and the town counted its gains, reckoned the loss its manners and morals had suffered, and returned to its normal quiet.

But its waking up had set its blood flowing faster and it has been a live town ever since, growing steadily, making public improvements, building houses and finding occupants faster than they could be put up, and all without any fictitious excitement.

It is for its coal mines—and these are half a dozen miles eastward—that Cañon City has the highest repute, however, and to these it owes much of its prosperity. They are the property of the ubiquitous Coal and Iron Company, who supply from here a large part of the fuel—and of the hardest and cleanest quality—used in the household all over the state. The railway consumes a great quantity too, for it has long been recognized as the best steam-maker. Two veins are now worked, one five feet and the other four feet in thickness. The mines are worked by slopes furnished with steam hoisters. The Coal Creek mine has been in operation for nine years. The company has recently opened two new slopes at Oak Creek, and are prepared to furnish 1,500 tons of coal per day from them. The total annual output of these mines is about 125,000 tons. The following are the analyses of these coals:

Water...	4.50	6.15
Volatile Matter.................................	34.20	36.03
Fixed Carbon...................................	56.80	52.82
Ash...	4.50	5 00
Total.......................................	100	100
Sulphur..		.65

Everybody will understand from this statement that this coal is worthless for coking, but most desirable as fuel. Our cook will swear to this, but declines to tell how much he stole at various times for culinary use.

Withal, Cañon City is a pretty town; one of the pleasantest places to live in in Colorado. Rows of large trees shade all the side-walks (and they are side-walks of planking, not mere gravel paths), and the ample spaces left about each house are filled with fruit trees, flowers and garden vegetables. To go into such a garden as one I visited in town is a surprise. A picturesquely built house, its adobe walls hidden by much climbing vinery, has its porch turned into a thickly-leaved bower by masses upon masses of clematis, whose white, thistly puffs of seed-down, each as large as a snow ball, are strung upon the green stem like monstrous beads. The garden, of which this cottage is the center, abounds in apple trees, pears, quince, plum, and peach trees, through whose spring blossoms thousands of bees go to and fro bearing burdens of honey to the neat store-houses under the shade. The lower part of the garden falls away, terrace fashion, to the river, and here are arbors of grapes, thickets of currant and gooseberry bushes, beds of asparagus, celery, and all sorts of good plants to make the pot-boiler happy. Down by the river stands a windmill, by which water is pumped to a reservoir, whence the whole garden and orchard can be irrigated and sprinkled.

This is only one of hundreds of gardens small and great where fruit and vegetables are raised for home use and for sale. Marvelous stories are told of the weight of the cabbages, of the girth of the beets, of the solidity of the turnips and strength of the onions that go hence. And as for apples, scores and scores of acres are being newly set out in apple trees, and almost square miles of "truck" fields will next year add their quota to the unsatisfied market. I was astonished when I saw how extensive and successful was the culture of fruit and garden sauce in and about Cañon City.

This comes from good soil and easy climate. They say some winters here are so mild that one hardly needs an overcoat at all; it must be remembered that though the elevation is high the latitude is low. I saw a field where clover had been cut three times a year for twelve years, yet showed no signs of running out; and as for alfalfa, they cut the crop quarterly.

The citizens think that their town is likely to prove a manufacturing center. I see no reason why it should not. The river falls there at a rate which furnishes a fine water power, already utilized to propel the public water-works. With the best of coal close by, and iron in abundance only a little way off, I should think the future would see machine shops and foundries placed at this point; while factories for woolen cloth, for making wooden-ware from pine and for various other industries adapted to the resources and market of the neighborhood, shoe factories and leather-work by machinery generally ought to be flourishing here some day, since hides ought to be tanned here instead of being sent wholly to the East.

In the State prison, which is situated here, there is already a shoe-factory, but most of the prisoners are engaged in quarrying and cutting stone. The quarries are in the side hill, and the stone is a yellowish sand-rock, very good for building. The fine-appearing prison-buildings and the lofty wall which encloses them are built of this stone, as can readily be seen from the car-windows. Much stone, in the rough and shaped, is shipped from these quarries to Denver and elsewhere, and the railway makes extensive use of it.

Just outside of the foot hills, where the sand stone is procured, are the "hog-backs"—elongated ridges of white lime-rock. These, also, are being leveled to supply the lime-kilns and also to be sent to Leadville, Argo and elsewhere for the use of the smelting furnaces as flux. Something like two hundred car-loads a week, I am told, go to Leadville alone; but the competition of lime ledges near Robinson and elsewhere north of Leadville is likely to diminish this shipment in future to that point. Possibly, though, if the newly discovered silver prospects over the hills near Blackburn turn out to be of any value, a home demand may make up for Leadville's discrepancies. Finally, petroleum seems to have been found here in quantities which will ultimately prove highly remunerative. Wells are being bored, and unexpectedly good results are obtained, so that high hopes are entertained that to her list of productions Colorado shall add in profitable quantities this wonderful substance—mineral oil—and the spirit of speculation and industry be given a new channel for its activity.

XVIII

IN THE WET MOUNTAIN VALLEY.

> For some were hung with arras green and blue,
> Showing a gaudy summer morn.
> . . . .
> And one a full-fed river winding slow
> By herds upon an endless plain.
>
> —TENNYSON.

AÑON CITY was by no means a bad place to stay, and we would have prolonged our visit to the benefit of our table, had not the railway yard been so busy a one that there was no rest for our cars, which were pulled about, here and there, by the necessities of train-forming, in a way we were far from enjoying, so we decided to go on. At the last minute, nevertheless, this happy-go-lucky crowd concluded that they were extremely anxious first to take a run over into the Wet Mountain valley. One gentleman, of uncertain influence, raised his voice against it, but was silenced so quickly it made his head swim. He had endeavored to point out that it would be more instructive to go down to the great coal mines, a few miles below; and far more fun to ascend Signal mountain and "see what we should see." He tried skillfully to arouse some enthusiasm by telling how, though it seemed within rifle-shot, it was really eighteen miles away; how it can be seen from the plains not only, but also from South Park and the peaks that surround; how, in consequence, the Utes chose it as one of their telegraph stations, and the early pioneers bound for Pike's Peak, saw from their camps the wavering smoke by day, or the signal fires at night, upon its summit, through which the Indians informed their companions of the invaders' movements. Thus it came to be known as Signal mountain, but in this gentleman's humble opinion the old Spanish name of "Pisgalo Peak" was better. All this was listened to with a sort of consolatory attention; nevertheless the speaker was compelled, not only to resign his plan, but to give orders otherwise.

Grown strong in the lap of the Wet Mountain valley, Grape creek assaults the red walls of rock that bar its progress to the Arkansas at the mouth of the Grand Cañon. The profusion of wild vines its waters nourish, makes its name a natural one, and they adorn its course as few streams in the West are garnished. These are particularly abundant along the rocky lower part of the stream, growing luxuriantly upon the arbors the great cottonwoods afford, and under the shelter of the warm, red walls, relieving the ruggedness of their abrupt slopes, "as if nature found she had done her work too roughly, and then veiled it with flowers and clinging vines."

"The entrance to Grape Creek cañon," writes an acquaintance, who

GRAND CAÑON OF THE ARKANSAS.

was there a little in advance of us, "for over a mile, follows the windings of the clear flowing creek, with gently sloping hills on either side covered with low spruce and piñon, and with grass plats and brilliant flowers in season far up their slopes, and the Spanish lance and bush cactus present their bristling points wherever a little soil affords them sustenance. . . About seven miles from the mouth of the creek a small branch cañon comes in from the right. It was once a deep cleft, with perpendicular sides, created in some convulsion of nature, but it has been gradually filled up with débris and broken rock until a sloping and not difficult path is made, by the sides of which a luxuriant vegetation has taken root, and the wild rose and clematis blooms with the humble blue-bell among the mossy bowlders. Climbing this path for a few hundred feet a side cleft is seen at the right, which seems to terminate in a solid wall. Following it to the breast, however, you find at the left a passage made by a water channel, with steps which ladies can easily pass with a little help, and we enter a narrow passage between high rocky walls. Turning again to the right, we follow this perhaps two hundred feet, and looking to the left we find before and above us the lofty arched dome of the "Temple." About twenty-five feet above where we are standing is a platform, perhaps fifty feet in width and six or eight feet in depth, over which projects far above the arching roof. Though the auditorium in front is rather narrow for a large audience, the platform is grand, and may be reached without great difficulty. Music sounds finely as it rolls down from the overhanging sounding-board of stone. From the platform deep cavernous recesses are seen at the sides, which time has wrought, but which are invisible from below. Moreover, the action of water slowly percolating through the back walls, carrying lime and spar in solution, has coated them with crystals, which gleam in sparkling beauty when the sunlight touches them early in the day. Farther up the cañon the rocks do not rise to so great heights, and the vista opens out into pleasant winding valleys well covered with grass, but there are several very interesting points where the action of internal convulsions upon the granite and syenite in elder ages, when they came hot from the crucible of nature, have rolled and twisted and kneaded the great rock-masses into most curious and notable shapes."

These beauties passed all too rapidly, the green expanse of the Wet Mountain valley opened before us. It seemed to merit its name, for the Sangre de Cristo, walling in its western side, was the abode of contending hosts of rain and snow, whose pale, dense phalanxes lent new sublimity to the noble battle-ground they had chosen; but the real "Wet Mountains,"—the old "Sierra Mojada" of the Spaniards, the "Green Horn" range of the dwellers on its eastern outlook—are the ragged range eastward.

The Wet Mountain valley has long been settled by ranchmen, and extensive herds pasture on its wide-sweeping hillsides. Grape creek, flowing from Promontory bluff and the hills to the southward, which separate this valley from Huerfano park and the drainage of the Cucharas, waters the center of the valley, and its banks are lined with meadows and farms. Each winter sees hay alone sent from these meadows to the value of not less than $150,000. Oats and barley, especially, do well, and most of the roots are grown successfully; very fine potatoes were transferred from those fields to our boiler, so that we have the best evidence of their excellence. The improved appearance of the numerous ranches, which in one or two places are agglomerated into hamlets, shows their prosperity, and the whole picture of the valley is one of the most pleasing in Colorado,—not only in point of natural beauty, but for its commercial and human interest, for Rosita is one of the oldest towns in Colorado.

"A legend runs," *vide* H. H., "that there was once another 'Little Rose,' a beautiful woman of Mexico, who had a Frenchman for a lover. When she died her lover lost his wits, and journeyed aimlessly away to the north; he rambled on and on till he came to this beautiful little nook, nestled among mountains, and overlooking a green valley a thousand feet below it. Here he exclaimed, 'Beautiful as Rosita!' and settled himself to live and die on the spot. A simpler and better authenticated explanation of the name is, that, when the miners first came, six years ago, into the gulches where the town of Rosita now lies, they found several fine springs of water, each spring in a thicket of wild roses. As they went to and fro from their huts to the springs, they found in the dainty blossoms a certain air of greeting, as of old inhabitants welcoming new-comers. It seemed no more than courteous that the town should be called after the name of the oldest and most aristocratic settler,—a kind of recognition which does not always result in so pleasing a name as Rosita (Tompkinsville, for instance, or Jenkins' Gulch). Little Rose, then, it became, and Little Rose it will remain."

But the metropolis of the valley, and the terminus of the railway at present is the newer town of Silver Cliff, a town which saw one of the "biggest booms" on record. The story goes that the first known discovery of silver here was in July, 1877, by the Edwards brothers, who had previously been running saw-mills on Texas and Grape creeks. Returning one warm evening from one of the mills to Rosita, Mr. R. J. Edwards stopped in the shade of a low bluff, jet-stained reddish rock, which stood out from the slope of a hill on the western side of the valley seven miles north of his destination. The peculiar appearance of the rock moving his curiosity, he procured an assay of it, when, to his astonishment, he was told that it ran twenty-four ounces in silver to the ton. In a few days the entire population of Rosita had migrated to the rock which they agreed to call the Silver Cliff, and were digging holes and testing for gold, since it was thought there was more of that than of the less valuable mineral to be obtained. But their efforts came to nothing; and as quick to be discouraged as they were to have their hopes aroused, the mercurial crowd vanished, and the black striped rocks enjoyed their previous solitude through all the next autumn and winter.

Then (this was in the spring of 1878) some sensible prospectors tried for silver and located the "Racine Boy" and various other properties right on the brow of the cliff, which have since proved of great value. This was the signal for a second rush, but the new comers, who dug holes everywhere and anywhere, like an immense colony of prairie badgers, each thought himself sure of millions, and held his bit of ground at so high a price that nobody would buy at all. This resulted in a panic, the effect of which was really for the prosperity of the critical camp, since capital now took hold and deep developments proceeded on some properties that had proved their worth.

It did not take long to evince the fact that ninety out of every hundred of the holes scattered so indiscriminately over the velvety knolls of Round mountain and the smooth, hard plain near by, were of no value; and also, on the other hand, to show enough paying mines to make it appear that the ore (at any rate that near the surface) all lay in a particular "belt," apparently culminating in the exposed ledges that had first attracted the miner's eyes.

The Hardscrabble mining district, in which both Silver Cliff and Rosita are situated, takes its name from a small creek that rises in the foothills on the west side of the Wet Mountain range, or Sierra Mojada, and, forcing its way through a wild and difficult cañon, flows into the Arkansas river seven or eight miles east of Cañon City. The mountains themselves are of red granite, which has been thrown up in the wildest confusion, and which the winds and rains in many places have carved into all sorts of fantastic shapes. The range is extremely rugged, almost destitute of large timber, and is impassable for wagons, except where roads have been built at great expense through the cañons and over the divide.

The western foothills of the Sierra Mojada generally present, at a distance, a smooth, rounded appearance, with now and then a ledge of rocks sticking out of the summit or side, and while on some of them timber of considerable size is growing, in most instances the vegetation consists solely of a few stunted evergreen bushes and a very thin growth of gramma grass. The soil on these hills is generally very thin, and, on approaching them, the surface is found to be covered with loose pieces of broken rock, which the frosts have detached and the rains washed out from beneath a slight covering of earth. It is in these foothills that all the best mines of the Hardscrabble district have been found.

The geological formation of this rich mineral belt is peculiar and very interesting. Resting upon and against the granite of the Wet Mountain range and its higher foothills, and extending down into the valley beyond the southern line of the belt, lies an enormous deposit of porphyry, or trachyte, a volcanic rock poured out and consolidated during the tertiary period. Its width is at least five miles, and its length is probably fifteen or twenty. Extending into the trachyte formation from the southwest, and following its general direction, is a tongue-shaped mass of granite about three-fourths of a mile wide and at least seven or eight miles long. When the trachyte was poured out this granite apparently formed a ridge which rose above the level of the fluid mass of the surrounding volcanic rock, and therefore was not covered by it. **That it does** not now stand higher than the surrounding country does not disprove this theory, because there are everywhere to be **found evidences** of terrible convulsions since the trachyte was deposited which have completely changed the face of this entire region. The mines here are found both in the granite and also in the trachyte. Winding through the porphyry, in a serpentine course, there is also a stream of obsidian, as it is called here, or volcanic glass, mixed with trachyte and quartz bowlders. This stream, where it has been examined, varies from a few feet to many rods in width, and in crevices of the bowlders which form the mass of it were found, on the Hecla claim, some very rich specimens of horn silver.

At Silver Cliff, and north of there especially, the trachyte rock has been shaken up and fractured in all directions, and in many places the crevices have been filled with iron and manganese, which has become oxidized, and with chloride of silver. This is the free milling ore which is found in all the mines that lie directly north of this town and adjoining it. The trachyte is of itself yellowish white; when it is stained with the black oxide of manganese and the red oxide of iron that variegates **the ores, it is sure to carry** silver, though this (in the form of a chloride) can rarely be seen. Sometimes, however, the silver can be seen upon the surface of a fracture in the form of a green scale, or appears in little globules of horn silver. While the rich ore is discovered in large masses, surrounded by leaner or less valuable rock, there is nowhere in the chloride belt anything that looks like a vein. The **rock** just covers the entire face of the country over an area two miles long and half a mile wide, and the whole mass of it contains at least a small quantity of silver.

The theory of the geologists, accepted by the miners, is, that the trachyte, after it became solidified, was shaken and broken up by some **great convulsion,** and that simultaneously, or afterward, silver, iron, **manganese,** and the other metals of which traces are found in the rock, **were disseminated** through crevices, either in water solutions or volatilized — in the form of gases. These solutions or gases are supposed to have come up through cracks in the earth's crust. Such a deposit is called in the old world "stockwork," and **Professor J. S. Newberry,** in writing recently of "The Origin and Classification of Ore Deposits," mentions **this as one of the t**wo most important examples of this kind of deposit that have come under his observation. The other is the gold deposit in Bingham cañon, **Utah.** None of the oldest miners ever saw before any ore that looked like this at Silver Cliff, and this explains their failure to discover its value until recently. The same is true of the quartzite gold ore in Bingham cañon. The miners worked for years there getting out silver-lead ores, but threw aside the gold ore as waste, **not dreaming of its value.**

But the mineral belt which I have described contains other classes of mines. At Rosita, in the "Pocahontas-Humboldt" lode, the trachyte, instead of being shattered and impregnated, has been rent asunder and a true fissure formed in it, filled with gray copper, galena, zinc blende, iron and copper pyrites and heavy spar—all carrying sulphide of silver. These form a narrow pay streak from one to eighteen inches wide, and

THE ROYAL GORGE

the remainder is filled with a gangue rock, generally of a trachytic formation. This vein extends for a long distance through the hills, and is inclosed by walls that are as clearly defined as those of a room. Other smaller veins of the same character have been found in the country north of Rosita, and on some of them valuable mines have been located and developed.

Still another class of mines in the same mineral belt remains to be mentioned. Those are what Professor Newberry has called the "mechanically filled" veins, and they include the "Bassick" and the "Bull-Domingo." The former is supposed to be a true fissure vein in the trachyte rock, the cavity of which, after the rocks were rent asunder, was filled with well rounded pebbles and bowlders, generally similar in constitution to the country rock. The interstices in this mass have been filled with tellurides of gold and silver, free gold, zinc blende, galena, and the pyrites of iron and copper carrying silver. These materials surround the stones in thin shells, the pebbles and bowlders forming nuclei about which the metallic substances crystallized. In the "Bull-Domingo," situated in the granite tongue, the stones are generally granite or syenite, and the cementing substance is argentiferous galena, which not only surrounds the stones, but in many cases entirely fills up the irregular spaces between them. In both of these cases it is supposed that the metallic matter came up from below in the form of a hot solution.

Silver Cliff has been a trifle disappointed, however. Not only were her streets laid out broad and straight, upon a splendid town site, over a considerably larger area than has yet been occupied, but two other towns, Westcliffe, where the railway station is, and Clifton, between the two towns, invite persons to buy town lots and build houses in rivalry. At present, however, Clifton's population consists chiefly of its town-agent, and there is one of the best opportunities to take your choice of building sites there that I know of in the Centennial state. Westcliffe has a big smelter, a hay-press, the water-works, and various other reasons for being a future village of importance.

Yet Silver Cliff is a fine town, and its streets are busy with miners and merchants and professional men, who know where their money is coming from and going to. The immense interests of the Silver Cliff Mining company, with its open, quarry-like mine, and its great mill, which has the reputation of being the finest in Colorado, employ a large number of men. Another mill, further down the creek, is running on the product of its mines, and a great deal of development-work along the whole belt is in progress, while prospects of rich strikes elsewhere keep things in a bright, hopeful condition.

As for Rosita, it was a thriving mining camp, half a dozen years before Silver Cliff and its chlorides were heard of. True fissure veins were disclosed, and a permanent town resulted, which is yet mining quietly but successfully, and making its people wealthy.

XIX

THE ROYAL GORGE.

High overarched, and echoing walls between.
—MILTON.

THE Grand Cañon of the Arkansas, and its culminating chasm, the Royal Gorge, lie between Salida and Cañon City, and form a sufficient theme for a chapter by themselves. It was on our return from Silver Cliff that we went there.

Situated only half a dozen miles west of Cañon City, the traveler going either to Leadville or Gunnison, begins to watch for the cañon as soon as he has passed the city limits, the penitentiary and the mineral springs. If he looks ahead he sees the vertically tilted, whitish strata of sandstone and limestone, which the upthrust of the interior mountains has set on edge, broken at a narrow portal through which the graceful river finds the first freedom of the plains,—becomes of age, so to speak, and commences, however awkwardly, that manly progress that by and by will enable it to take its important place in the commerce of the world,—

"—— The river,
Which through continents pushes its pathway forever,
To fling its fond heart in the sea."

Running the gauntlet of these scraggy warders of the castle of the mountain-gods within, the train boldly assaults the gates of the castle itself. From the smoothness of the outer world, where the eye can range in wide vision, taking in the profiles of countless noble chains and lowlier but serviceable ridges; where the sun shines broadly, and its light and heat are reflected in shimmering volumes from expanses of whitened soil, the eager traveler now finds himself locked between precipitous hillsides, strewn with jagged fragments, as though the Titans had tossed in here the chips from their workshop of the world. He strives for language large enough to picture the heights that with ceaselessly growing altitude hasten to meet him. He searches his fancy after images and similitudes that shall help him comprehend and recall the swiftly crowding forms of Nature's massive architecture. He taxes his eyes and mind and memory to see and preserve, until he can have leisure to study this exhibition of the depth and breadth of the barrier that so long has loomed before him in silent majesty, yet for which the world has found no better name than the Rocky mountains. He has gone past it,—gone over it, it may be; now he is going *through* it. The track, as he rushes ahead, seems bodily to sink deeper and deeper into the earth, as though the apparent progress forward only resulted in impotent struggles to keep from sinking deeper, like an exhausted swimmer in swift waters. The roar of the yeasty, nebulous-green river at his side, mingles with the crashing echoes of the train, reverberating heavenward through rocks that rise perpendicularly to unmeasured heights. The car is stunned, and the mind refuses to sanction what the senses report to it.

Then a new surprise and almost terror comes. The train rolls round a long curve, close under a wall of black and banded granite, be-

BROWN'S CAÑON.

side which the ponderous locomotive shrinks to a mere dot, as if swinging on some pivot in the heart of the mountain, or captured by a centripetal force that would never resign its grasp. Almost a whole circle is accomplished, and the grand amphitheatrical sweep of the wall shows no break in its smooth and zenith-cutting façade. Will the journey end here? Is it a mistake that this crevice goes *through* the range? Does not all this mad water gush from some powerful spring, or boil out of a subterranean channel impenetrable to us?

No, it opens. Resisting centripetal, centrifugal force claims the train, and it breaks away at a tangent past the edge or round the corner of the great black wall which compelled its detour, and that of the river before it. Now what glories of rock-piling confront the wide-distended eye. How those sharp-edged cliffs, standing with upright heads that play at hand-ball with the clouds, **alternate** with one **another, so that** first the right, then the **left, then the right** one beyond strike on our view, each one half obscured by its fellow in front, each showing itself level-browed with its comrades as we come even with it, each a score of hundreds of dizzy **feet in** height, rising perpendicular from the water and the track, splintered atop into airy pinnacles, braced behind against the almost continental mass through which the chasm has been cleft.

This is the Royal Gorge!

But how faintly I tell it—how inexpressible are the wonders of plutonic force it commemorates, how magnificent the pose and self-sustained majesty of its walls, how stupendous the height as we look up, the depth if we were to gaze timidly down, how splendid **the massive shadows at** the base of the interlocking headlands,—**the glint of** sunlight on the upper rim and the high polish of the crowning points! One must catch it all as an impression on the retina of his mind's **eye,**—must memorize it instantly **and** ponder it afterward. It is ineffable, but the thought of it remains through years and years a legacy of vivid recollection and delight, and you never cease to be proud that you have seen it.

There is **more cañon after** that—miles and miles of it—the Grand Cañon of the Arkansas. In and out of all the bends and elbows, gingerly round the promontories whose very feet the river laves, rapidly across the small, sheltered nooks, where soil has been drifted and a few adventurous trees have grown, **noisily** through the echoing cuttings, the train rushes westward letting **you down** gradually from the tense excitement of the great chasm, **to the** cedar strewn ledges that fade out into the the gravel bars and the **park-like spaces of the open valley** beyond Cotopaxi.

Thomas Paine tells us in his *Age of Reason*: "**The sublime and the ridiculous** are often so nearly related, that it is **difficult to class** them separately." It is good philosophy also, that the higher the strain the longer the rebound; so no excuse is needed for asking you **to** enjoy **as** heartily as we did, the story an old fellow told us at the supper station, who dropped the hint that he had been one of the "boys" who had **helped** push the railway through this cañon. Moreover, **he** helped us to a new phase of human nature as exemplified in the mind of an "old timer."

The influence of the **cañon** on the ordinary tourist, perhaps, will be comparatively transient, fading into a dream-like memory of amazing mental impressions. Not so with the man who has dwelt, untutored, for many years, amid these stupendous hills and abysmal gorges. His imagination, once aroused and enlarged, continues to expand; his fiction, once created, hardens into **fact;** his veracity, once elongated, stretches on and on forever. Of all natural curiosities he **is** the most curious,— more marvelous than even the Grand **Cañon** itself.

Strictly sane and truthful in the day-time, he speaks only of commonplace things; but when the night comes, and the huge mountains group themselves around his camp-fire like a circle of black Cyclopean tents, he shades his face from the blaze and bids his imagination stalk forth with Titanic strides. Then, if his hearers are in sympathy, with self-repressed and nonchalant gravity, he pours forth in copious detail his strange experiences with bears and bronchos, Indians and serpents, footpads and gamblers, mines and mules, tornadoes and forest-fires. He never for a moment weakens the effect of his story by giving way to gush and enthusiasm; he makes his facts eloquent, and then relates them in the careless monotone of one who is superior to emotion under any circumstances.

We could not find our old-timer in these most favorable circumstances, but ensconced behind

> "Sublime tobacco! which from east to west,
> Cheers the tar's labors, or the Turkman's rest,"

he seized his opportunity in our discussion of the heroic engineering by which the *penetralia* of the Royal Gorge was opened to the locomotive, and began:

"Talk about blastin'! *The boy's yarn about blowin' up a mountain's nothin' but a squib to what we did when we blasted the Ryo Grand railroad through the Royal Gorge.

"One day the boss sez to me, sez he, 'Hyar, you, do you know how to handle gunpowder?'

"Sez I, 'You bet!'

"Sez he, 'Do you see that ere ledge a thousand feet above us, stickin' out like a hat-brim?' Sez I, 'You bet I do.'

"'Wall,' sez he, 'that 'll smash a train into a grease-spot some day, ef we don't blast it off.'

"'Jess so,' sez I.

"Wall, we went up a gulch, and clum the mountain an' come to the prissipass, and got down on all fours, an' looked down straight three thousand feet. The river down there looked like a lariat a' runnin' after a broncho. I began to feel like a kite a' sailin' in the air like. Forty church steeples in one war'n't nowhar to that ere pinnacle in the clouds. An' after a wHile it begun rainin' an' snowin' an' hailin' an' thundrin' an' doin' a reglar tornado biznis down thar, an' a reglar summer day whar we wuz on top. Wall, there 'wuz a crevice from where we wuz, an' we sorter slid down into it, to within fifty feet o' the ledge, an' then they let me down on the ledge with a rope an' drill. When I got down thar, I looked up an' sez to the boss, 'Boss, how are ye goin' to get that 'cussion powder down?' Yer see, we used this ere powder as 'll burn

---

*If anybody doubts the full veracity of this tale, he is referred to Colonel Nat. Babcock, of Gunnison City.

like a pine-knot 'thout explodin', but if yer happen to drop it, it 'll blow yer into next week 'fore ye kin wink yer eye.

"'Wall,' sez the boss, sez he, 'hyar's fifty pound, an' yer must ketch it.'

"'Ketch it,' sez I. 'Hain't ye gettin' a little keerless—s'pose I miss it?' I sez.

"'But ye must n't miss it,' sez he. ''T seems to me yer gettin' mighty keerful of yourself all to wunst.'

"Sez I, 'Boss, haul me up. I'm a fool, but not an idgit. Haul me up. I'm not so much afeared of the blowin' up ez of the comin' down. If I should miss comin' onto this ledge, thar's nobody a thousan' feet below thar to ketch me, an' I might get drownded in the Arkansaw, for I kain't swim.'

"So they hauled me up, an' let three other fellers down, an' the boss discharged me, an' I sot down sorter behind a rock, an' tole 'em they'd soon have a fust-class funeral, and might need me for pall-bearer.

"Wall, them fellers ketched the dynamite all right, and put 'er in, an' lit their fuse, but afore they could haul 'em up she went off. Great guns! 'T was wuss 'n forty thousan' Fourth o' Julys. A million coyotes an' tin pans an' horns an' gongs ain't a sarcumstance. Th' hull gorge fur ten mile bellered, an' bellered, an' kep' on bellerin' wuss 'n a corral o' Texas bulls. I foun' myself on my back a lookin' up, an' th' las' thing I seed wuz two o' them fellers a' whirlin' clean over the mountain, two thousan' feet above. One of 'em had my jack-knife an' tobacker, but 't was no use cryin'. 'T was a good jack-knife, though; I do n't keer so much fur the tobacker. He slung suthin' at me as he went over, but it did n't come nowhar near, 'n' I don't know yet what it was. When we all kinder come to, the boss looked at his watch, 'n' tole us all to witness that the fellers was blown up just at noon, an' was only entitled to half a day's wages, an' quit 'thout notice. When we got courage to peep over an' look down, we found that the hat-brim was n't busted off at all; the hull thing was only a squib. But we noticed that a rock ez big ez a good-sized cabin, hed loosened, an' hed rolled down on top of it. While we sat lookin' at it, boss sez, sez he,

"'Did you fellers see mor'n two go up?'

"'No,' sez we, an' pretty soon we heern t' other feller a' hollerin', 'Come down 'n get me out!'

"Gents, you may have what's left of my old shoe, if the ledge had n't split open a leetle, 'n' that chap fell into the crack, 'n' the big rock rolled onto the ledge an' sorter gently held him thar. He war n't hurt a har. We wer n't slow about gettin' down. We jist tied a rope to a pint o' rock an' slid. But you may hang me for a chipmuck ef we could git any whar near him, an' it was skeery business a foolin' roun' on that ere verandy. 'T war n't much bigger 'n a hay-rack, an' a thousan' foot up. We hed some crowbars, but boss got a leetle excited, an' perty soon bent every one on 'em tryin' to prize off that bowlder that 'd weigh a hundred ton like. Then agin we wuz all on it, fer it kivered th' hull ledge, 'n' whar 'd we ben ef he 'd prized it off? All the while the chap kep' a hollerin', 'Hurry up; pass me some tobacker!' Oh, it was the pitterfulest cry you ever heern, an' we didn't know what to do till he yelled, 'I'm a losin' time; hain't you goin' to git me out?' Sez boss, 'I 've bent all the crowbars, an' we can't git you out.'

"'Got any dynamite powder?' sez the feller.

"'Yes.'

"'Well, then, why 'n the name of the Denver 'n' Ryo Grand don't you blast me out,' sez he.

"'We can't blast you out,' sez boss, 'fer dynamite busts down, an' it 'll blow you down the canyon.'

"'Well, then,' sez he, 'one o' ye swing down under the ledge, an' put a shot in whar it's cracked below.'

"'You're wiser 'n a woman,' sez boss. 'I 'd never thought o' that.'

"So the boss took a rope, 'n' we swung him down, 'n' he put in a shot, 'n' was goin' to light the fuse, when the feller inside smelt the match.

"'Heve ye tumbled to my racket?' sez he.

"'You bet we have, feller priz'ner!' sez the boss.

"'Touch 'er off!' sez the feller.

"'All right,' sez boss.

"'Hold on!' yells the feller as wuz inside.

"'What's the racket now?' sez the boss.

"'You hain't got the sense of a blind mule,' sez he. 'Do you s'pose I want to drop down the canyon when the shot busts? Pass in a rope through the crack, 'n' I'll tie it 'roun' me, 'n' then you can touch 'er off kind o' easy like.'

"Wall, that struck us all as a pious idea. That feller knowed more 'n a dozen blind mules—sed mules were n't fer off, neither. Wall, we passed in the rope, 'n' when we pulled boss up, he guv me 'tother end 'n' tole me to hole on tighter 'n' a puppy to a root. I tuck the rope, wrapped it 'round me 'n' climb up fifty feet to a pint o' rock right under 'nuther pint 'bout a hundred feet higher, that kinder hung over the pint whar I wuz. Boss 'n' t'other fellers skedaddled up the crevice 'n' hid.

"Purty soon suthin' happened. I can't describe it, gents. The hull canyon wuz full o' blue blazes, flyin' rocks 'n' loose volcanoes. Both sides o' the gorge, two thousan' feet straight up, seemed to touch tops 'n' then swing open. I wuz sort o' dazed 'n' blinded, 'n' felt ez if the prisipasses 'n' the mountains wuz all on a tangle-foot drunk, staggerin' like. The rope tightened 'round my stummick, 'n' I seized onto it tight, 'n' yelled:

TWIN LAKES

"'Hole on, pard, I'll draw you up! Cheer up, my hearty,' sez I, 'cheer up! Jes az soon 'z I git my footin', I'll bring ye to terry firmy!'

"Ye see, I wuz sort of confused 'n' blinded by the smoke 'n' dust, 'n' hed a queer feelin', like a spider a swingin' an' a whirlin' on a har. At last I got so'z I could see, 'n' looked down to see if the feller wuz a swingin' clar of the rocks, but I could n't see him. The ledge wuz blown clean off, 'n' the canyon seemed 'bout three thousan' feet deep. My stummick began to hurt me dreadful, 'n' I squirmed 'round 'n' looked up, 'n' durn my breeches, gents, ef I was n't within ten foot of the top of the gorge, 'n' the feller ez wuz blasted out wuz a haulin' on me up.

"Sez I when he got me to the top, sez I, 'Which eend of this rope wuz *you* on, my friend?'

"'I dunno,' sez he. 'Which eend wuz *you* on?'

"'I dunno,' sez I.

"An', gents, to this day we can't tell ef it was which or 'tother ez wuz blasted out."

It was afternoon and we were weary—sated—with sublimity; so we ran straight away to Leadville, and left until our return an examination of the Arkansas Valley.

XX

THE ARKANSAS VALLEY.

And he gave it for his opinion, that whoever could make two ears of corn or two blades of grass, to grow upon a spot of ground where only one grew before, would deserve better of mankind, and do more essential service to his country, than the whole race of politicians put together.—JONATHAN SWIFT.

HE interest of the Grand Cañon of the Arkansas, though it culminates between the narrow walls of Royal Gorge, by no means ceases there. For many miles after, immense piles of rocks are heaped on each side, great crags frown down, and the river comes tumbling to meet you down a series of green and white cataracts. The walls are highly colored, and the whole scene exceedingly interesting. Toward the western end there is a break in the gorge, through which fine pictures of the Sangre de Cristo peaks present themselves close by, and then the rocks are heaped up again into the grand defile of Brown's cañon, where one of our illustrations was made.

Just before entering Brown's cañon, a branch road can be seen running off to the northward. That is the short road up to Calumet, where the Colorado Coal and Iron Company have iron mines of great value and in constant operation, for the ore is suitable for the making of Bessemer steel. These mines are open, quarry-like excavations, and the ore is therefore more easily handled than is usual. The grade on this branch, four hundred and six feet to the mile, is said to be the heaviest in the world where no cog-wheels are used. Only a few empty cars can be hauled up; and the difficulty is almost as great in descending, for it requires at least four cars, dragging with hard-set brakes, to hold an engine under control in going down. Marble and lumber in great quantities are also shipped down this little branch from the neighborhood of Calumet.

Passing some hot mineral springs, where are bathing arrangements, near the head of Brown's cañon, the train runs into the busy yard at Salida. This town was formerly South Arkansas, and I surprise the Madame by telling her that no longer ago than 1874, I pitched a tent where it now stands upon ground which had no vestige of civilization near it. Salida is a Spanish word, meaning a junction, and is applicable in two ways. It is at the confluence of the Arkansas with its large branch from the south, and it is the junction of the northern system of railway which we are following to Leadville and beyond, with the main line going west from here to Utah and California. It is therefore a lively railway center,—the end of divisions, the headquarters of round-houses, repairing shops, etc. Besides, it is rapidly growing, and increasing in importance as a busy mercantile center.

The valley of the Arkansas north of Salida, we see as we go on again, nourishes much agriculture, which continues to be seen—at least

THE OLD ROUTE TO LEADVILLE.

in the shape of hay ranches—as far as Riverside, the first station above Buena Vista. There Mr. Leonhardy has seven miles, more or less, under cultivation, and carries on a highly profitable farm. His extensive hay-barns are close to the track, and his horse-mowers show how scientifically it is cut. All the cereals are grown there, or at any rate have been grown; but wheat, though it becomes very plump and hard, has so precariously brief a season in which to mature, that it is not profitable, and hence no great amount is now planted. Of oats, rye and barley, however, hundreds of acres are cut annually, yielding in each case above the average number of bushels to the acre of eastern crops. I have seen some very fine samples of all these grains, which, of course, find abundant sale close at home, and hence are unheralded outside.

Then in the way of "roots," large plantations are made, and fine results brought about. Potatoes are particularly successful, one hundred and fifty bushels, or about fifteen thousand pounds weight, being the ordinary crop expected to the acre; turnips, beets, onions, etc., doing equally well in their way. The only things that can not be produced here, in fact, are such tender plants as melons, squashes, cucumbers, and the like. Even these may often be brought to maturity if their beginnings are nurtured under glass, but as a matter of regular gardening, they are not considered profitable.

Apart from this locality not much farming is visible, except close to Salida, where the road runs over the top of a dry mesa,—one of the terraces into which the former river has cut the glacial gravels of the valley-margin. Down in the lower "bottoms," where irrigation is very easy, one sees some miles of continuous fields cultivated in hay and grain. The close clusters of ranch-buildings, the stacks of straw, the yellow and green squares of stubble and the black threads of the dividing fences, with the diminutive dots of men moving to and fro with wagons, recall the prairie states. We also note the number of cattle seen all along the lower part of the valley,—and the cheapness and excellence of the beef we bought in all this part of Colorado.

Buena Vista is a town of considerable size and seeming solidity, which is prettily placed among the cottonwoods. These give a name to the stream not only, but to the expansion of the valley, which is known as Cottonwood park. The supply point not only for the Chalk Creek mines on Mt. Princeton, but for the remoter settlements on the other side of the range, Buena Vista seems to have a good chance for long life. One sees here the big, trailed wagons in all their glory, and the voice of the burro is heard in the land, complaining of his burdens and bewailing the lost friskiness of his unfettered youth.

Below Granite we pass through almost a cañon. The inclined and splintered rocks of reddish granite and gneiss rise very high at certain points on the eastern bank of the river, and the water itself is in continual ebullition among large bowlders, falling meanwhile at such a grade that the track cannot follow it, but must needs rise away above it. The scene here is one of extreme desolation. There is nothing *pretty* in the whole landscape short of the small snow-banks that remind us of scattered sheep browsing on the crest of the range. Almost the only relief to the sterility—sterile not only in respect to pleasing vegetation, but in any comfortable suggestiveness—is when the sun shines suddenly straight down some rift-like gulch in the precipitous walls, transmuting what seemed a crystal-clear atmosphere into a golden dust finer than any flakes that ever came out of the gravels.

Now we are rapidly approaching Granite, a town twenty-five years old; and presently we catch sight of the great gold placers that formerly made the fame of this locality. They are still operated in a quiet, scientific method, and one large flume crosses the track at a height of fully fifty feet. The western bank has been ploughed up by water and turned topsy-turvy over a long area, exposing its innermost pebbles and bowlders, all well cleaned and white by their second scrubbing.*

Three miles west of Granite lie the charming "Twin Lakes," but we are frustrated in our attempt to reach them on the only day we wished to spare for that purpose.

During all the summer, carriages from the lake meet passenger

---

* If the reader cares to know more about the lively times that used to occur now and then in Granite, years ago, he can find some incidents in my "Knocking 'Round the Rockies" (New York, Harper & Brothers, 1882), on page 70 and following.

trains at Twin Lakes station, four miles above Granite, in order to carry visitors to this lake.

"Of all the health and pleasure resorts of the upper Arkansas Valley," I have read, "the Twin Lakes are perhaps the most noted. Water is nowhere too plentiful in Colorado, the largest rivers being usually

THE SHAFT HOUSE.

narrow and rapid streams, that seldom form an important feature in the extended landscapes, and these lakes are all the more prized for constituting an exception. They are fourteen miles south of Leadville. The larger of the two lakes is two and one-half miles in length by one and one-half in width, and the other about half that size. The greatest depth is seventy-five feet. These lakes possess peculiar merits as a place of resort. Lying at an altitude of 9,357 feet,—over one and three-fourths miles,—at the mouth of a cañon, in a little nook, surrounded by lofty mountains, from whose never-failing snows their waters are fed, their seclusion invites the tired denizens of dusty cities to fly from debilitating heat and the turmoil of traffic, to a quiet haven where Jack Frost makes himself at home in July and August. On the lakes are numerous sail and row boats, and fishing tackle can always be obtained. Both lakes are well stocked with fish, and the neighboring streams also abound in mountain trout. Surrounding the lakes are large forests of pine, that

THE TWIN LAKES.

add their characteristic odor to the air. The nearest mountains, whose forms are reflected in the placid waters, are Mount Elbert, 14,351 feet in height, La Plata, 14,311,—each higher than Pike's Peak,— Lake Mountain, and the Twin Peaks. Right royal neighbors are these. And across the narrow Arkansas valley rises Mount Sheridan, far above timberline, flanked by the hoary summits of the Park range. The hotel and boarding-house accommodations are good, and will be rapidly extended. During the summer months there is an almost constant round of church and society picnics and private pleasure parties coming down to the lakes from Leadville, so that nearly every day brings a fresh influx of visitors, enlivening the resort, and dispelling all tendency to monotony.

"Twin Lakes is the highest of all the popular Rocky Mountain resorts, and furnishes an unfailing antidote for hot weather. Even in midsummer flannels are necessary articles of apparel, and thick woolen blankets are indispensable at night."

There are mines in the mountains back of Twin Lakes, and gradually a permanent settlement is growing up there, which is reached all the year round by stage from Leadville. This stage passes over Hunter's pass, and carries the mail to some important camps on the other side of the range,— Independence, Highland, et cetera. The main point of interest, I hear, is at Independence, which is said to be much such a camp as Kokomo, and standing at a greater altitude than even Leadville. The veins are true fissures filled with quartz containing free gold, iron and copper pyrites. The Farwell Mining company are the chief operators, and have recently erected what has been pronounced

BOTTOM OF THE SHAFT.

the finest stamp-mill in Colorado. It consists of thirty stamps, and cost $2.87½ cents a hundred pounds for carriage from the railway to its site. This feat required the building and repair of roads to an extent that has been of immense benefit to the public. Besides this mill there

ATHWART AN INCLINE.

is an old one of twenty stamps, and additions are to be made. A few miles further on, is the flourishing camp of Aspen, standing in a beautiful valley 7,500 feet above the sea. This is the locality of the Smuggler mine owned by Mackay and other Eastern capitalists. It is described as "a large lead of fine-grained galena, carrying native silver in wire form." Aspen is a good type of the "magic" town, where lots increase a thousand per cent. in value in six months.

This brings us to Malta, a station in the midst of a wide waste of denuded gravel, where we turn up California gulch to Leadville, bidding good-bye, for a little, to the white crests of the Saguache range,—Harvard, Yale, Princeton, Antlers and others that have been our constant companions. Turn where we will in this region, we can not long escape the sight of snow-smothered peaks. It is impossible to get away from them. This river valley is a great basin surrounded on all sides by mountains that hasten to bid winter welcome before summer has thought of saying farewell to the valleys. As in that wonderful story, wherein we are constantly reminded that "the villain still pursued her," so here the mountains unceasingly confront us, and every changing mood can be studied by our eager eyes.

Malta seems to be a great place for charcoal, many groups of the white conical ovens being visible on the blackened and denuded side-hills.

Charcoal is an extremely important element in smelting operations, and enormous quantities are made, to the destruction of all the forests, so that the burners have to go farther and farther with their ovens, or else, as most of them are doing, have wood brought from increasing distances. A favorite method is to build a flume and float the timber, in short pieces, down from the higher woods; or else, simply to make a trough, laying it partly on the ground and partly on trestles, so as to secure proper levelness. It is great fun to watch them shuteing wood or ties (for the "tie-punchers" adopt the same expedient) down the slope of the high, steep hills. Little choice is made in the kind of wood burned.

The effect of these charcoal makers is very plain as we climb up the devious track through the hills of California gulch to Leadville.

The trees were cut which once stood dense over the whole of the gulch, and then every vestige of brushwood, grass,—everything was burned away, so that the ash-strewn soil and the charred stumps alone remain of the former verdancy. Into this oddly desolate tract the town has pushed itself without altering it much for the better. The outer

THE JIG DRILL.

suburbs of a town are seldom pleasing, and Leadville is no exception. The burned stumps, thick as the original forest, give a general black aspect to the whole scene. Fences are few, and amount to the merest pretense of enclosures, more than an unbarked pole or two, strung along the boundary, being rare. The streets are mere spaces, for there is no difference at all between the outside and the inside of the fence. The public highway finds itself as best it can among the stumps, and the householder rarely bothers himself to pull one out of his front yard.

This is not mere rough neglect, and, in the center of town, of course does not exist. It shows that the citizens, as a rule, do not care to make fine their surroundings, because they have not come to stay. They are a generation of pilgrims, even though, under endless protest, they may linger, or be held here, all their lives, and be buried in the stony little graveyard, under the yellow fumes of the smelters down the creek, at the last. Inside, though, the houses glow with pretty things and abound in luxuries. Here, men combat the outward roughness and resolve that they will be comfortable in compensation for the inclemency outside.

And so we come to Leadville, the "Camp of the Carbonates."

XXI

THE CAMP OF THE CARBONATES.

> Moored in the rifted rock,
> Proof to the tempest's shock,
> Firmer he roots him the ruder it blow.
>
> —SCOTT.

F the men who sprang from the stones Deucalion cast behind him set themselves to make homes, the result must have been a close counterpart of Leadville, Colorado." Such was the phrase with which the present writer began an article upon the "Camp of the Carbonates," printed in *Scribner's Monthly* for October, 1879. Though the Leadville of to-day has graduated from the overgrown mining camp it then was, into a pretentious city of twenty thousand people, and boasts all the "improvements," yet the interest connected with the town, for the world at large, is chiefly historical.

Historical! Why, Leadville is only seven years' old now; but the years have been eventful, and history is made fast in this state.

The site of Leadville has a pre-historic interest also,—almost mythological in fact, for have not five and twenty years crept by since then. This is the well verified tradition:

"After the rush to Pike's Peak, in 1859, which was disappointing enough to the majority of prospectors, a number of men pushed westward. One party made their way through Ute pass into the grand meadows of South Park, and crossing, pressed on to the Arkansas valley, up which they proceeded, searching unsuccessfully for gold, until they reached a wide plateau on the right bank, where a beautiful little stream came down. Following this nearly to its source, along what they called California gulch, they were delighted to find placers of gold. This was in the midsummer of 1860; and before the close of the hot weather, ten thousand people had emigrated to the Arkansas, and $2,500,000 had been washed out, one of the original explorers taking twenty-nine pounds of gold away with him in the fall, besides selling for $500 a 'worked out' claim from which $15,000 was taken within the next three months. Now this same 'exhausted' gravel is being washed a third or fourth time with profit.

"The settlement consisted of one long street only, and houses, even of logs, were so few that the camp was known as 'Boughtown,' everybody abandoning the wickyups in winter, when the placers could not be worked, and retreating to Denver. During the summer, however, Boughtown witnessed some lively scenes. One day a stranger came riding up the street on a gallop, splashing mud everywhere, only to be unceremoniously halted by a rough looking customer who covered him with a revolver and said:

"'Hold on there, stranger! When ye go through this yere town, go slow so folks can take a look at ye!'

"No money circulated there; gold-dust served all the purposes of trade, and every merchant, saloon-keeper and gambler had his scales. The phrase was not 'Cash up,' but 'Down with your Dust,' and when a man's buck-skin wallet was empty, he knew where to fill it again. It was not long, however, before the placers were all staked off, and the claims began to be exhausted. Then the town so dwindled that, in half a dozen years, only a score were left of the turbulent multitude, who, in '60 and '61, made the gulch noisy with magical gains and unheeded loss. Among the last of their acts was to pull down the old log gambling hall, and to pan two thousand dollars out of the dirt where the gamblers had dropped the coveted gains. This done everybody moved elsewhere, and the frightened game returned to thread the aspen groves and drink at the once more translucent streams of California gulch, where eight millions of dollars had been sifted from the pebbles.

"One striking feature of this old placer-bar had impressed itself unpleasantly upon all the gold-seekers. In the bottoms of their pans and rockers, at each washing there accumulated a black sand so heavy that it interfered with the proper settling of the gold, and so abundant that it clogged the riffles. Who first determined this obnoxious black sand to be carbonate of lead is uncertain. It is said that it was assayed in 1866, but not found valuable enough to pay transportation to Denver, then the nearest point at which it could be smelted. One of the most productive mines now operated is said to have been discovered in '67, and in this way: Mr. Long, at that time the most poverty-stricken of prospectors, went out to shoot his breakfast, and brought down a deer; in its dying struggles the animal kicked up earth which appeared so promising that Long and his partner Derry located a claim on the spot. The Camp Bird, Rock Lode, La Plata and others were opened simultaneously outside the placers, but all these were worked for gold, and though even then it seemed to have been understood in a vague way that the lead ores were impregnated with silver, nobody profited by the information Thus years passed, and I and many another campaigner in that grand solitude, riding over those verdant slopes, passing beneath those somber pine woods, camped, hunted, even mined at what now is Leadville, and never suspected the wealth we trampled upon.

"Among the few men who happened to be in the region in 1877, was A. B. Wood, a shrewd, practical man, who, finding a large quantity of the heavy black sand, tested it anew and extracted a large proportion of

FREMONT PASS

silver. He confided in Mr. William H. Stevens, and they together began searching for the source of this sand-drift, and decided it must be between the limestone outcropping down the gulch and the porphyry which composed the summit of the mountain. Sinking trial shafts they sought the silver mean. It took time and money, and the few placer-washers there laughed at them for a pair of fools; but the men said nothing, and in the course of a few weeks they 'struck it.' Then came a period of excitement and particularly lively times for the originators of the enterprise. Mr. Stevens was a citizen of Detroit, and finding a chance for abundant results from labor, but no laborers wherewith to 'make the riffle,' he went back to Detroit and persuaded several scores of adventurous men to come out here and amuse themselves with carbonates.

"They came, hilariously, no doubt, with high anticipations of sudden wealth and the fulfilling of wide ambitions; came to find the snow deep upon the ground, and winter bravely entrenched among the gray cliffs of Mosquito and the Saguache. No one could work; every one was tantalized and miserable; discontent reigned. It was the old story of Baker and the San Juan silver fields. They took Wood and Stevens, imprisoned them in a cabin, and even went so far toward the suggestion of hanging as to noose the rope around their necks. At this critical moment, reprieve came in the shape of a capitalist who appeased the hungry crowd with cash and stayed their purpose until the weather moderated and digging could be begun.

"As spring advanced and the mountains became passable, there began a rush into the camp, for the report of this wonderful rejuvenation of the old district had spread far and wide. The Denver newspapers took up the laudation of the region. The railways approaching nearest, advertised the camp all over the East for the sake of patronage; and many an energetic prospector, and greedy saloon-keeper, and many a business man who wanted to profit by the excitement, started for Leadville. It was early spring; the snow lay deep on the lofty main range of the Rocky mountains which had to be crossed, and filled the treacherous passes, but the impatient emigrants could not wait. To be first into Leadville was the aim and ambition of hundreds of excited men, and to accomplish this, human life was endangered and mule flesh recklessly sacrificed. Companies were organized, who put on six-horse stages from Denver, Cañon City and Colorado Springs, and ran three or four coaches together, yet private conveyances took even more than the stages, and hundreds walked, braving the midwinter horrors of Mosquito pass.

"Meanwhile an almost continuous procession of mule and ox trains were striving to haul across that frightful hundred miles of mountains the food, machinery and furniture which the new settlement so sorely needed, and which it seemed so impossible to supply. Ten cents and more a pound was charged for freight, and prices ranged correspondingly high, with an exorbitant profit added. Hay, for example, reached $200 per ton.

"Nor were all who came rough or even hardy characters. There were among them men of wealth and brains, young graduates of colleges eager for a business opening, engineers and surveyors, lawyers, doctors, and a thousand soft-handed triflers who hoped to make a living in some undefined way out of the general excitement. Many of these gentlemen went to stay and took their wives, or, more usually, waited until they had prepared some sort of a home, and then sent for them. What stories some of these ladies tell of their stage-journey through those wintry mountains! How many wagons, heavily loaded with freight, did they see overturned by the roadside! How many dead mules and horses did they count! How many snow-banks did they fall through! how many precipices escape! how many upsettings avoid by the merest margin of consummate good driving! I knew of three ladies who for twenty-four hours were packed in a stage with a lot of drunken men, who could only be kept within the bounds of decorum and safety by being sung to sleep. The driver was utterly powerless to control them, and had as much as he could do to steer his six horses over that icy road. The crazy men said, 'Sing to us, we like it, and if you don't we'll dump you into the snow!' and sing they did, all night long. Whether this incident be considered laughable or pathetic, it is literally true. In the summer the stage passenger was not frozen, but was choked to slow death by impenetrable clouds of dust, and in the seasons between he was engulfed in mud. Verily that hundred miles of staging at fifteen cents a mile, with only thirty pounds of baggage allowed free, was the Purgatory of Leadville, and helped wonderfully to make one contented with his reception.

"With the beginning of 1879, the steady current that had flagged somewhat during the tempestuous last months of 1878, burst into a perfect freshet of travel. Log huts, board shanties, canvas tents, kennels dug into the side hill and roofed with earth and pine boughs, were filled to repletion with men and women, and still proved insufficient to shield the eager immigrants from the arctic air and pitiless storms of this plateau in the high Sierras. Men were glad to pay for the privilege of spreading their overcoats or blankets on the floor of a saloon and sleeping in stale smoke and the fumes of bad whisky—an atmosphere where the sooty oil lamps burned with a weak and yellow flame. Perhaps the dice rattled on till morning above the sleepers' heads, the monotonous call-song of the dealers lulling them to an unquiet doze in the murky air, only to be awakened by the loud profanity of some brawler or sent cowering under the blankets to escape the too free pistol-balls that fly across the billiard table. Even the sawdust floors of these reeking bar-rooms were not spacious enough to hold the two hundred persons a day who rushed into Leadville, and every dry-goods box upon the curbstone, every pile of hay-bales in the alley, became a bedroom for some belated traveler.

"But the era of saloon-floors and empty barrels did not last long. Enterprising men built huge hotels, and opened restaurants and great lodging-tents and barracks; strangers joined in twos and threes, cut logs and planted cabins as thick as corn. . . Every day chronicled some new accession of wealth, some additional tapping of the silver deposits which were firmly believed to underlie every square foot of the region. It seemed all a matter of luck, too, and skilled prospecting found itself at fault. The spots old miners had passed by as worthless, 'tenderfeet' from Ohio dug down upon, and showed to be rich in 'mineral.' One of the first mines opened — the Camp Bird — was discovered by the Gallagher brothers, two utterly poor Irishmen. Another early piece of good fortune was that of Fryer, from whom Fryer Hill, one of the most productive districts, derives its name. He lived in a squatty little cabin on the side-hill, where the dirt floor had become as hilly as a model of the main range, and the

CASCADES OF THE BLUE

rough stone fire-place in the corner was hardly fit to fry a rasher of bacon; but one day he dug a hole up near the top of the hill, hiding himself among the secret pines, saying nothing to anybody, and a few yards below the surface struck a mine which has already yielded millions of dollars without being urged. Innumerable incidents might be related of the patience and expense and hardship which resulted in failure; of the equal pluck and endurance that brought success; of happy chance or perfect accident divulging a fortune at the most unexpected point. The miners have a proverb, 'Nobody can see into the ground,' and the gamblers an adage, 'The only thing sure about luck is that it's bound to change!'

"One of the grimmest of these tales is that attached to the Dead Man claim, which is briefly as follows: It was winter. Scotty had died, and the boys, wanting to give him a right smart of a burial, hired a man for twenty dollars to dig a grave through ten feet of snow and six feet of hard ground. Meanwhile, Scotty was stuffed into a snow bank. Nothing was heard of the grave-digger for three days, and the boys, going out to see what had happened to him, found him in a hole which, begun as a grave, proved to be a sixty-ounce mine. The *quasi* sexton refused to yield, and was not hard pushed, for Scotty was forgotten and staid in the snow-bank till the April sun searched him out, the boys meanwhile sinking prospect-holes in his intended cemetery.

"One mine had its shaft down one hundred and thirty-five feet and the indications of success were good. Some capitalists proposed to purchase an interest in it, and a half of the mine was offered them for $10,000, if taken before five o'clock. At half-past four, rich silver ore was struck, and when at half-past five the tardy men of money came leisurely up and signified their consent to the bargain, the manager pointed at the clock, and quietly remarked:

"'The price of a half interest in this mine now, gentlemen, is sixty thousand dollars.'

"Prospectors went everywhere seeking for carbonates, radiating from this center up all the gulches, and over the foot-hills, delving almost everywhere at a venture. One day, at a hitherto unheard-of point, wealth comes up by the bucketful out of the deep narrow hole, that has been pierced so unostentatiously. Instantly the transformation begins, and the lately green hill-side, refreshing to the townsman's eye, becomes forlorn in its ragged exposure of rock and soil where the forest has been swept away, while trial-mines grow as thickly upon its surface as pits on the rind of a strawberry. All these young mines, good or bad, looked much alike, and were equally inaccessible and unkempt. There were no roads, hardly any wagon-tracks and few paths. Every man went across lots, the shortest way, pushing through the remnant of the woods, clambering over the prostrate trunks and discarded tree-tops, whose straight trunks had been felled and dragged away to the saw-mill, or chopped into six-foot lengths for posts and logging. Teams must go around, but life was too short for the man afoot to follow them; holding his painful breath, he scaled straight up the steep and slippery ascent.

"But it is time to say something of the processes of getting out the ores, and perhaps the best way is at once to attack the geological structure of the region.

"Leadville appears to lie upon the eastern edge of the lava area of the state. The last of the trachyte peaks are at the head of Mosquito pass. Underneath the camp, and on all the hills where her riches are stored, the soil is found to be a porphyritic overflow overlying a highly silicified dolomite, that goes by the common name of 'limestone.' Between these two formations (*i. e.*, under the porphyry and above the dolomite) are found the mineral beds. Various theories have been advanced as to the reason for their position, so novel in the experience of silver mining, and some of the explanations are a burlesque of geology, though uttered in dead earnest. Those who are best qualified to decide, although confessing limited observation, suggest what seems to me the simplest theory and the one nearest the truth. The mineral constituents of the ores are carbonate of lead in large quantity, silica, oxides of iron and manganese, and the precious chloride of silver. Sometimes the lead occurs as a sulphide, and there are some other insignificant components. Now it is possible that the original constituent parts of all these minerals should be contained in a porphyritic eruption. Deposits of galena and some other minerals are now occasionally found buried in the porphyry, or occupying slender fissure-veins through it. Moreover, all these minerals are capable of solution in water charged with carbonic acid, which, of course, was present in abundance, and the suggestion is that they have leached downward through the porphyry until they struck the limestone floor, which became in time so highly silicified, as to admit no further penetration of water, whereupon the valuable deposits that we are now prying out gradually accumulated. The silicified surface of the lime, and the semi-saturated line of the porphyry, next the carbonate, are known as the 'contacts;' and when the miners strike this, they have good cause to be hopeful of near success. The presence of great beds of kaolin (hydrated silicate of alumina), derived from the thorough decomposition of porphyry or granite, or both together, and the presence of hydrate of magnesia with beds of semi-opal (always an aqueous production), argue in favor of the truth of this explanation.

"The general fact of this position of the ores being understood, let me suppose that our prospectors have been more than ordinarily successful; that they have dug not more than a hundred feet, have curbed their shaft securely with timber, have struck the greenish-white porphyry, and finally have met with the longed-for 'contact,' which separates the mineral bearing rock from the barren gangue. They have been little troubled by water, and they have done all their work with the help of one man, and the ordinary windlass. There being every indication that wealth is just beneath their picks, they erect over the shaft a frame-work of heavy timbers, called a 'gallows,' and hang in it a large pulley. A little at one side, close to the ground, is fixed a second pulley. Under this, and over the upper one is reeved the bucket-rope, and a mule is hired to walk away with it, when the bucket is to be drawn up, creeping back when the bucket goes down. This is a 'whip.' The next advance in machinery is the 'whim,' which consists of the same arrangement of gallows and pulleys as before; but instead of a mule walking straight out and back, the mule travels round and round a huge revolving drum, that carries the hoisting-rope. If you care to go down one of these shafts you may stand in the bucket, or you may unhook it, and, placing your foot in a noose, be lowered away in the bucket's place. If your head is strong there is no great danger.

"When the miner really 'strikes it,' and the brown, crumbling, ill-looking ore begins to fill the bucket to the exclusion of all else, assaying fifty or a hundred or four hundred ounces to the ton, a house is built over the shaft, and a steam-engine supersedes the patient mule.

"The depth at which a mine may be found (if at all) can hardly even be guessed at. Paying 'mineral' has been met with from the surface to more than three hundred and fifty feet in depth. Usually the shafts are over a hundred feet deep.

"The deposit having been tapped, digging out the ore begins. This is done by means of horizontal passage-ways or tunnels, known as 'drifts,' which are driven into the rock from the bottom of the shaft.

"As the ores are brought to the surface they are scanned by an experienced person, and the best pieces thrown in a heap by themselves, while the ordinary ore is cast upon the 'dump' or pile which accumulates at the mouth of the mine, and makes a little ruddy terrace on the green or snowy hill-side. From this dump wagons haul the ore away to be sold, the best part often being put in hundred-pound sacks, about as large as quarter-barrel flour-bags, before being sold. Very rich ore is likely to be bought by regular purchasers, who either have them smelted in Leadville or forward them to smelting-works at Pueblo, Denver, St. Louis, and Eastern cities. The inferior grades are sold by the ton to some one of the dozen smelters here in town, the price being governed by the market quotations of silver in New York on the day of the sale, less several deductions amounting in all to about twenty-five per cent. as the reducer's margin for profit, and plus three to five cents per pound for all the lead above twenty-one per cent. which the ore carries. Silver and gold are estimated in ounces; lead and copper in percentages; but allowance is not made for both of the latter metals in the same ore. The ore is hauled to the smelting-works by four or six-mule teams, for the most part, the driver not sitting on the wagon, but riding the nigh wheeler, guiding his team by a single very strong rein which goes to the bits of the leaders, and handling the brake by another strap. He is in the position of a steersman in the middle of his craft, and his 'bridge' is the saddle. Every load is set upon the scales, recorded, and then shoveled into its proper bin. A thin-faced, dusty-haired youth leaned half asleep against a shady corner at one of these mills, recording the tons and fractions of a ton in each load as he lazily adjusted the balance. His air was of one so utterly listless and bored that I was moved to remark cheerily as I went by:

"'You haven't chosen the most exciting part of this business.'

"'No,' he answered dryly, while an indescribable twinkle came into his carbonated countenance. 'No, but I'm trying to do my duty. You know the poet says, "They also serve who only stand and weigh it."'

"That fellow had a history, but I haven't time to tell it. Leadville is full of such characters, and it only needs to put one's self *en rapport* with their happy-go-lucky good humor and stoicism under all sorts of fortune to find these miners, at heart, the best fellows in the world. They have a high regard for a gentleman, but a hatred of a swell; no objection to good clothes, but a horror of 'frills;' a high respect for genuine virtue, but boundless hatred of cant; an admiration for nerve amounting to worship, but a contempt of braggadocio that often results in an impulsive puncturing of both the braggart and his boasts. A 'tender-foot,' that is, a new arrival from the East, green in the ways of mountain life, they consider fair game for tricks and chaff. Usually they attempt to frighten him, and his behavior at such initiatory moments determines, to a large extent, his future standing in the camp.

"But this is a digression from the subject in hand, which is the reduction of the ores. The smelters cannot be allowed to cool off, and so are run the twenty-four hours through. One evening we made up a party and visited one of the great smelters. Its chimney-stacks pour noxious smoke over a nest of cabins down on the bank of the creek, and guide us, by scent as well as sight, through the streets and across the vacant lots. The broad upper floor is divided along one side into a series of bins, opening outwardly into a shed, under which the teams drive that bring the ore. Each owner's lot is put into a bin and kept separate until sampled and paid for. This sampling is a process akin to homeopathy. Supposing one hundred tons are to be sold at the smelter. Every tenth ton, as fast as delivered, is set aside to be sampled. This ten tons is then subdivided,—perhaps by being carried from one part of the floor to the other in wheelbarrows,—every tenth load being set aside. The single ton thus remaining contains many large, hard lumps. These are roughly screened out and put through a crusher, which chews them into fragments no larger than walnuts. The heap of a ton of broken material thus formed is now separated in a very ingenious way, by

MOUNT OF THE HOLY CROSS.

catching a few lumps of the ore from each shovelful in a 'scoop,' which a man holds above the wheelbarrow wherein the main portion is carted back to the original pile in the bin. The saved portion, which has happened to fall into the scoop, constitutes a new sample, to be further reduced, by successive crushings and screenings, until finally there remains only a pound of earth as the perfect representation of the average quality of the five hundred tons of rocky ore offered from the mine. This pound is then ground to powder on the bucking-board, and a tenth or twentieth is taken for the scientific fire-test, or 'assay,' which shall determine its value. All these processes go on at night as well as by day.

"The red-brown ores lay in little heaps about the floor when we entered, divided from one another by low partitions. Men with spidery wheelbarrows were cruising about, dumping a pile of precious earth here, shoveling up another there, with seemingly aimless purposes, and the bins were only like so many openings to a mine, so deep were the shadows hiding their recesses. Across the room, lanterns showed four great circular chambers of iron, from whose depths hoarse rumblings drowned in a deep, steady bass the energetic crunch-crunch of the insatiate ore-chewers. Wide door-ways admitted into these dungeons, where surging volumes of murky vapors were confined, and through their hot portals red-shirted men hurled the raw material that should be digested, and the worthy part of which should issue from the furnaces below in a bright and costly stream: first a barrow load of carbonate ore, next one of charcoal, then a third of iron and limestone-flux.

"Day after day, night after night, these monsters are fed with this diet, varied in proportions according to the richness and metallurgical qualities of the ore that is being smelted. It requires very good judgment to determine just how much foreign material and lime is needed to produce the best results with the constantly varying ores. Luck may find the silver ore but science must extract the bullion. Most profit accrues to the smelter when the ore produces from seventy-five to two hundred ounces of silver, and contains a goodly proportion of iron and lead.

"Leaving the dungeons, we pick our way down the slope of a small mountain of ore, and enter below, where the engine and boilers throb, and the openings at the bottom of the furnaces give exit to the silver and the slag we saw shoveled in above as ore. And what an exit! The low roof shuts down close and dark upon the huge black cylinder of iron and bricks that holds in its heart the molten metal. There are pipes and valves, and draft-ways, and beams and braces, but they show indistinct in the gloom, and are nothing beside that great central mass, begrimed with soot and the dust of arsenic and oxides of lead. Watch that workman. He lifts a lance and stepping near the base of the furnace, where a single spark directs his aim, gives two or three quick thrusts. How mighty an effect the simple act evokes! The gloomy and ghost-haunted chamber becomes a home of fire; the grim furnace breathes out gaseous flames of blue and green, with tongues of light which hover playfully over a cataract of melted red metal bubbling, spouting, plunging out of that Plutonic throat and falling in hissing streams into the iron bowl waiting to catch its hot flood. The little lady who is with us, seeing the sparks fly, draws timidly outside the doorway and none too soon, for without warning the whole place becomes volcanic. No longer a steady stream of artificial lava rolls down the iron channel, but the liquid metal bursts its bounds and becomes a fountain. The furnace is hidden in lurid gases out of which spring volley upon volley of burning fragments that scatter showers of fire over the whole foreground.

"The slag-pot is a conical vessel, with a rounded apex, poised, base uppermost, on four little legs; when it is full, an iron frame work of a cart runs up, seizes it on opposite sides as though with two hands, and wheels it, glowing and fuming, out where a mole of slag is pushing itself over into the white gravel of the gulch, and where it is deposited red and crackling among heaps of like cones, some fading into the ashy hues of spent heat, some black and shining like inverted crucibles of polished iron. It was an uncanny vision: the huge rough outlines of the great mill, with its high chimneys and beacons of flame and smoke; the blaze within, the wan moonlight outside, and the sinewy men with skeleton carts leaping about in the glare of the spouting slag, handling shapely burdens of fiery refuse.

"While the worthless slag is doing so much sputtering and making so lively a show of itself, the silver and lead have quietly sunk to the bottom as fast as the heat liberated them from the mass of the boiling ore, and now come oozing up from a small exit far below the slag-spout, into a well at the side of the furnace. As fast as needful, this liquid 'bullion' is ladled out and poured into iron moulds, where it remains until it cools into solid 'pigs' or bars of lead weighing about fifty pounds each, and carrying about two per cent. of silver. These pigs, when cool, are stamped with the smelter's name and the number of the car-load to which they will belong. Then from each one is cut a fragment, and these pieces—when the whole 'run' of the furnace has been made—are collected and re-cast and assayed to determine the value and selling-price of the bullion."

The foregoing paragraphs, culled without indicating the omissions, and so, perhaps reading abruptly, it must be remembered, were written in the early summer of 1879. Yet, to a great degree, the picture outlined in that (now old) magazine article holds good to-day. There are many more people here, and the coming of the Denver and Rio Grande railway has brought the world nearer and multiplied the means of trade. It has reduced prices, afforded ready transportation out and in, civilized the town. Harrison avenue has become a metropolitan street, crowded with fine business houses, where you can buy almost as many things as in Denver, and the hills in the outskirts are crowded with more mine-houses and riddled with more tunnels than formerly. But all this is an advance in degree, not an addition of a new kind. The paving of the central streets, the erection of large business buildings, the introduction of public water and gas, the police, the fire-patrol, the morning and evening papers, the telephone and what not, are all indications of the thrift and prosperity of the people but render the city less characteristic and peculiar. The Leadville of '79 **in which we took** a keen interest is now a thing of the past.

After dinner, the Madame and I go up as of yore, to a cottage **we wot of** that commands a pleasant view, and sit watching the night put the shading into the picture. But I tell her it is not the picture I used to see and enjoy. That was a great map of new, bare houses spread out before us, seemingly without arrangement or form. **The** steady drone of late planing mills and the subdued, eager rasp of steam-**saws** begrudging the approach of darkness, told how grew the magic **town that** was overrunning the plateau, exploring the gulches, and swarming up the flanks of the half-cleared foot hills. It was a town **without** high buildings or towers, church-spires or foliage. In **the** clearness with which every detail is seen at a great distance, the houses looked smaller than they really were. It was all rough and ragged, **yet** all the more picturesque.

Slowly the **long, sober twilight deepens in the** valley into gloaming, and sinks thence into a gloom out of which, one by one, peep the lights. Still, outlines are not lost, and the massive figures of the foothills thrust themselves hugely through the veil that night is dropping, **solid** and blue and forbidding. It is a picture of perfect sweetness and peace,—a poetic picture in which one can imagine nothing that is harsh, **or selfish, or mean.** And overhead the mountains tower, rank behind rank, peak crowding peak, the pinnacles vying in being the last to hold **the** lingering rays of the sun, whose light now enkindles the heights until all **the** wide snow-fields burn rosily. Then **one by one the glittering** banks fade into the softest of ash-tints **as the reluctant sun bows** itself away, and the shadows of **the** blackening ridges fall athwart the arctic panorama that fills the horizon. Keeping pace, the lights **of the city increase, shining duskily through a** purple haze of **smoke** and mist. Clearer above **this ethereal** stratum of haze, gleam the jewel-points that show where huge **engines** are tirelessly at work, and where prospectors and campers have built their fires on the hill-sides, and sit about them boiling their coffee and gossiping on the events of the day **and** the prospects of the **morrow.** Then the Madame and I saunter homeward—for our comfortable cars **seem** very homelike to us these frosty evenings—breathing the resinous **flavor of** the crisply fragrant **spruce, and** watching the stars spring hastily over the coruscant line that **traces the** serrated crest of the snowy range.

Leadville at night is a scene of wild hilarity, and yet of **remarkable order.** The omnipresent six-shooters that used to outnumber the men of **a mining camp ten years ago are** rarely seen **here in** public. If men **carry pistols, it is in their pockets;** and the shoot-the-lights out ruffianism **of the old** frontier days rarely shows even a symptom of revival. You find a city **of** twenty thousand people or so within the limits and

MARSHALL PASS—EASTERN SLOPE.

up the sides of the hills that overlook the town, where hundreds of mine-houses, spouting ceaseless jets of steam from ever-laboring engines, and hundreds of dumps of earth and ore brought to sudden daylight from their beds in the heart of the hill, tell the story of Leadville's prosperity. The rough old camp has crystallized into the city she resolved to become.

As for these mines—what shall be said. Fryer Hill, which was the source of Leadville's "boom," has gone into obscurity under the newer glory of its rivals, Carbonate and Breece hills. It is said that Fryer Hill proved a great collection of "pockets," very rich so long as they lasted, but liable at any time to be exhausted. The other hills, however, seem not to have suffered the geological turmoil through which Fryer passed, and, therefore, when a deposit of ore is struck, one may be reasonably sure of its holding out as long as any one man or generation of men would be likely to feel an earthly interest in its development. Men now know pretty well, or think they do, what ones of the hundreds of "discovery shafts" sunk are really worth continuing, and there is a constant tendency to the consolidation of adjacent properties into the hands of large companies controlling vast capital, and pushing operations with quiet dignity. The bullion product of Leadville increases year by year, and gives an annual output varying from $17,000,000 to $19,000,000.

The yard of the Denver and Rio Grande railway, where our cars lay for a whole week, is a scene of never ceasing activity. This is the terminus not only of the main line from the east and south, but also of two branches, one down the Blue river and the other over to the Eagle River valley. Both have to cross the continental range, and abound in scenery so picturesque that, in the phrase of the penny-a-liner, "to be appreciated must be seen." That being the case, we propose to "see" it.

XXII

ACROSS THE TENNESSEE AND FREMONT'S PASSES.

> 'Unto the towne of Walfingham
> 'The way is hard for to be gon;
> 'And verry crooked are those pathes
> 'For you to find out all **alone**.'
> —Percy's Reliques.

ACCORDING to the virtuous intention of the last paragraph, we went one day over to Red Cliff and the Eagle river. The branch of the railway which runs thither, leaves the main line at Malta, and takes in some very pretty scenery.

From Malta the line skirts the wide hay-meadows between the village and the Arkansas river; I saw men spreading manure there, too, and was told they had raised oats successfully. The whole mouth of California gulch, here, is a vast bed of clean, drifted gravel, the result of the gold hydraulic operations above, the placers having been worked more or less continuously for twenty years.

Rising along a tortuous path cut at a heavy grade, as usual, into the side hills, we mount slowly into Tennessee Pass, which feeds the head of the Eagle river on one side and one source of the Arkansas on the other. It is a comparatively low and easy pass, covered everywhere with dense timber, and a wagon-road has long been followed through it. There was nothing to be seen except an occasional pile of ties, or a charcoal oven, save that now and then a gap in the hills showed the gray rough summits of Galena, Homestake, and the other hights that guard the Holy Cross. At each end of the Pass is a little open glade or "park," where settlers have placed their cabins and fenced off a few acres of level ground whereon to cut hay, for nothing else will grow at this great elevation.

One of the side-valleys, coming down to the track at right angles from the southwestward—I think it is Homestake gulch—leads the eye up through a glorious alpine avenue to where the cathedral crest of a noble peak pierces the sky. It is a summit that would attract the eye anywhere,—its feet hidden in verdurous hills, guarded by knightly crags, half-buried in seething clouds, its helmet vertical, frowning, plumed with gleaming snow,—

> "Ay, every inch a king."

It is the Mount of the Holy Cross, bearing the sacred symbol in such heroic characters as dwarf all human graving, and set on the pinnacle of the world as though in sign of possession forever. The Jesuits went hand in hand with the *Chevalier Dubois*, proclaiming Christian gospel in the northern forests; the Puritan brought his Testament to New England, the Spanish banners of victory on the golden shores of the Pacific were upheld by the fiery zeal of the friars of San Francisco; the frozen Alaskan cliffs resounded to the chanting of the monks of St. Peter and St. Paul. On every side the virgin continent was taken in the name of Christ, and with all the *eclat* of religious conquest. Yet from ages unnumbered before any of them, centuries oblivious in the mystery of past time, the Cross had been planted here. As a prophecy during unmeasured generations, as a sign of glorious fulfillment during nineteen centuries, from always and to eternity a reminder of our fealty to Heaven, this divine seal has been set upon our proudest eminence. What matters it whether we write "God" in the Constitution of the United States, when here in the sight of all men is inscribed this marvelous testimony to his sovereignty! Shining grandly out of the pure ether, and above all turbulence of earthly clouds, it says: Humble thyself, O man! Measure thy fiery works at their true insignificance. Uncover thy head and acknowledge thy weakness. Forget not, that as high above thy gilded spires gleams the splendor of this ever-living Cross, so are My thoughts above thy thoughts, and My ways above thy ways.

Red Cliff is a bright, fresh little camp, made of sweet-smelling, new lumber just out of the saw-mill; it looks *spruce* in a most literal sense. Perhaps a thousand or fifteen hundred persons live in and about there, though you will not see a quarter of that number except on Saturday nights and Sundays. The hotel where I stopped was made of canvas, but they gave me a good meal, and when bed-time came took me off to another tent-roofed shanty, which I occupied all to myself, surrounded by feminine finery and knicknacks, from tooth-powder and hair-pins to ruffled skirts and a sewing stand; however, the window-curtains consisted of two very "loud" copies of the *Police Gazette*, so I locked my door with extra care for fear the fair owners might unexpectedly return.

The mines in the neighborhood of Red Cliff—if you saw the toppling piles of rust-stained quartzite which hung over the gulch, you would not need to ask why the name was given—are of varied character, and of wide reputation.

Discovered only in 1878, it was at once seen that here in Battle mountain were enormous deposits of carbonate of lead carrying silver, which was so free from any refractory elements, like zinc or antimony, and so abundant in lead, that they were unexcelled in the world for the purposes of smelting. It has always been a drawback in the Lead-

MARSHALL PASS—WESTERN SLOPE.

ville ores that they contained lead in too small a proportion to the silver, copper and other constituents, to make straight smelting feasible; that is it is necessary to mix into each charge an addition of "flux,"—chiefly lead, in order properly to perform the operation of smelting. This Red Cliff ore, however, is so rich in lead, frequently running sixty, seventy or eighty per cent., that no accessory is needful, and it "smelts itself," as they say. In consequence, the carbonates of this district are in great demand at Leadville, and really bring more than their intrinsic value, since the smelters are anxious to get them to mix with the more refractory home product, and so get enough lead in the charge to secure the silver of both kinds of ore. Most of the ores from this camp, therefore, are shipped to Leadville; and not only that, but a large quantity of the bullion made here is sold there also and re-melted in order to furnish the necessary lead.

Here, as well as further down the river, some streaks of gold-quartz are found, and a stamp-mill is about to be erected. Fissure veins of silver ore are also known and worked somewhat, and much is expected of this branch of production in future. But thus far the chief reliance of the district is placed upon the carbonate ores of silver. You will find all the hills and granite ledges and quartzite overflows about here punched full of prospect-pits; but it is only on the southern slope of Battle mountain that mines worth mention have been developed as yet. "The whole interior of Battle mountain," one who knew said to me, "seems to be one bed of carbonate of lead and silver." Then he took me into the sheds of his smelter and showed me bin after bin full of brick-red, and rust-brown and dark and bright yellow earth, which lay in crumbling pieces like dried mud, or had fallen into mere sand, and told me that that was the general style of the ore. I lifted a handful and it was as heavy as shot: no doubt about that being lead. This stuff is almost too easy to mine; it is like digging into a sand-bank, and every foot of the way must be carefully protected by a timber tunnel to prevent its caving in. A man can pull down three or four tons a day, to ship, and it is only requisite to wheel it to the brow of the steep hill-side at the mouth of the mine, and hurl it down a shute a thousand feet or so to the railway track in the cañon.

This cañon of the Eagle, through which the railway runs, offers one of the keenest pleasures in Colorado to the lover of scenery, and one of the points of pilgrimage to the disciple of trout-fishing. The limpid green waters of the pretty river, fed, just here, by Turkey creek bringing the melted snows of the main range, and by the Homestake coming from the foot of the Holy Cross, dash with laughter and gurgle through a narrow defile of gayly colored rocks and thence pour out to rest awhile in the parks before its struggle with Elbow Cañon down below. From here to the mouth of the river, it is between fifty and sixty miles according to the line of the railway, which will, some day, closely follow its banks down the Grand to **Grand Junction**. The elevation is uniformly so great, even after you get fairly out of the mountains, that agriculture is hopeless, excepting the cultivation of some of the hardiest vegetables, like turnips, and perhaps risky crops of oats and barley.

At the mouth of the Roaring Fork of the Grand (which is just below where the Eagle debouches), some remarkable mineral springs bubble out of the ground. These have long been held in high esteem by the Indians and hunters, and now a little settlement has grown up around them called **Glenwood.** A hotel, bath houses and other facilities for a pleasant and healthful time have been erected, and the place is likely to prove a favorite summer resort. Many men are living and digging upon the headwaters of Brush creek, Gypsum creek, and other tributaries. Just below, where the Eagle river discharges itself, the Grand receives the Roaring Fork and various other pretty large tributaries, so that it becomes a noble stream by the time the great **Gunnison** reinforces it, and it mingles its waters with the Green river, which has come all the way from the National Yellowstone Park, to make the mighty Rio Colorado.

Hither will come the painters, who need not go to Switzerland for snowy bergs, nor to Scotland for lochs, nor to Norway for splendid forests of pine and spruce. No mountains I know of abound in more that is picturesque; but it is always some phase of the *grand* rather than the *pretty.* The scenery is wild and savage and primeval, being the stock of which beautiful landscapes are made, rather than the culture that gentler airs and more temperate winds bring upon the face of the earth nearer the sea and the equator. The naturalist also may come here with profit. The fauna and flora are boreal and western. The geologist and mineralogist and meteorologist will find much here to interest them, and clear up doubtful points.

This splendid, hilly, well-timbered, well-pastured, well-watered western edge of the state, is the grandest hunting-ground in the United States, and it will be long before the bears and mountain lions and wild cats; the wolves and foxes; the mountain-sheep, the elk, the two deers and the antelope, are driven from its shady courts and disappear from the wide and sunny ranges. Long let us say, in fond hope, if not in serious expectation, that *never* shall the dread word *exterminated* be written after the name of any of the wild animals whose utility as game or for beauty of form makes them of interest to us.

Another excursion from Leadville was out on the stub of a line to be extended down the Blue river toward Middle Park.

To reach the valley of Blue river "the range" must, of course, be crossed. The line from Leadville follows up the Arkansas and reveals to us how small are the beginnings of great things in the way of water-courses; how a miserable, shallow, wiggling little runlet, which you can dam with a couple of shovels of mud and stand astride of like another Colossus of Rhodes, may push its way along, undermining what it cannot overthrow; sliding around the obstacle that deemed itself impassable; losing itself in willowy bogs, tumbling headlong over the error of a precipice or getting heedlessly entrapped in a confined cañon; escaping down a gorge with indescribable turmoil, and always

CRESTED BUTTE MOUNTAIN AND LAKE.

growing bigger, bigger, broader and stronger, deeper and more dignified; till it can leave the mountains and strike boldly across a thousand miles of untracked plain to "fling its proud heart into the sea." Hark! what does it prattle up here where we can leap its ripplings, and the red willows tangle their blossoms and shade it from side to side?—

"Clear and cool, clear and cool,
By laughing shallow and dreaming pool,
Cool and clear, cool and clear,
By shining shingle and foaming weir."

Listen again below, where it rushes triumphant from the adamantine gates that sought to imprison it:—

"Strong and free, strong and free,
The flood gates are open away to the sea;
Free and strong, free and strong,
Claiming my streams as I hurry along,
To the golden sands and the leaping bar
And the taintless tide that awaits me afar."

Almost in the very springs of the river, where an amphitheatre of gray quartzite peaks stand like stiffened silver-gray curtains between the Atlantic and the Pacific, we curl round a perfect shepherd's crook of a curve, and then climb its straight staff to the summit of Fremont's,— the highest railway pass in the world. The pathway is so hidden in great woods, and the grim giants of the Mosquito range are still so inaccessibly far above you, even when you have reached the sterile oberland, above the trees, that you hardly realize the fact that you are 11,540 feet —considerably over two vertical miles—above the sea.

Once more on the Pacific slope, with the crossing of this range, we see the first trickling of Ten Mile creek, and enter the edge of one of the famous mining districts of the state, catching a sidelong glimpse of the Holy Cross as we descend.

"Although its now well-known silver mines," says a recent historical account, "are of recent date, the district is not a new one, having been run over by gold hunters in the 'flush times' of California gulch, Buckskin Joe and other famous gold-camps of early days. Gold was found in the bed of Ten-Mile creek, and in the connecting gulches, . . . among them McNulty's gulch, said to have yielded more gold in proportion to its size than any other workings in the state, and many fine nuggets of unusual size were taken from it. . . . The discovery in 1878, of the famous Robinson group of mines, followed, by the White Quail and Wheel of Fortune discoveries, attracted large numbers of prospectors to the new camp, and in spite of the ten feet of snow that covered the ground during the winter of 1878-'79 locations were made, and shafts and tunnels begun in every direction. During the winter the town of Carbonateville was settled, and for a time promised to become a thriving camp. On the 8th of February, the town of Kokomo, which, with its younger rival, Robinson, is now a prosperous and growing mining camp, with two smelters in operation, was located. In the spring of 1880 Robinson's camp began to build up rapidly, under the support of the great Robinson mines, and the fostering care of the late Lieutenant Governor Robinson, and soon became a formidable rival to Kokomo. The many discoveries made during the spring and summer of 1880, brought the district into a prominence second only to that of Leadville, and a large amount of capital was invested in the development of its many promising mines and prospects. Two smelters were erected at Kokomo, and one near the old town of Carbonateville, while extensive works, consisting of furnaces, roasters, etc., were put up at Robinson to work the ores of the Robinson mine. A railroad to connect the district with Leadville on the south and Georgetown on the east, was projected, and partially graded during the summer, but was finally absorbed by the enterprising managers of the Denver and Rio Grande Company, who, with a watchful eye for the future, began the construction, under the name of Blue River extension of the Denver and Rio

Grande, of a road, which in spite of the many and great difficulties encountered, was completed to Robinson on the 1st of January, 1881. Much of the grading and most of the track-laying were done under a heavy fall of snow, the range being crossed in midwinter, affording a striking instance of the energy and contempt of obstacles characteristic of Western railroad builders."

The Robinson mines alluded to, now abandoned so far as productive work is concerned, and generally considered a failure, were called the best mining property in the state only a year ago. They were discovered in 1878, the ore proving to be chiefly galena and iron, with large pockets of rich oxidized ore,—the "mud carbonates" so-called. A year later this mine passed into the possession of a stock company, headed by the late Lieut. Governor George B. Robinson, with a capital of $10,000,000. Extensive and thoroughly constructed tunnels, etc., were begun, which were soon interrupted by litigations out of which grew a small war. In the course of this Governor Robinson was accidentally shot by a guard, in November of 1880. These troubles settled, ore began to be produced in large quantities until the winter of 1882, when work suddenly ceased, the stock of the company fell to nothing, and the report was given out that the mine was a failure.

RUBY FALLS.

Moving on down the pleasant valley, whose level bottom is carbonate tinted, not with ore dust, but with an almost continuous thicket of stunted red willows, we pass the Chalk Mountain mines, the Carbonate Hill district, Clinton gulch, where gold ore is alleged to be worth more attention than it is receiving, and so come to Elk mountain and Kokomo, a locality which has had a wonderful history. In the fall of 1880 she had only the "White Quail" mine as a steady producer. A little later the "Aftermath" group came to the front. Now probably not less than fifteen distinct mining claims on Elk mountain are making a steady output of ore. This ore is a hard carbonate, running about twenty-five ounces in silver and twenty-five per cent. in lead, besides a third of an ounce in gold, which is carefully separated at the smelter. Much of it is so admirably constituted that it "smelts itself,"—that is, it requires little or no addition of lead, iron and other accessories to its proper fluxion.

We were told of alluring pictures of mountains and cañons below Kokomo; of timber-belts and pleasant uplands; of green meadows and sparkling streams beloved of trout and bass, and the drinking places of deer in the twilight. But our plan would not permit us to go on to Dillon, the present terminus, much less beyond it. Instead, we must turn back, make a swift run down the Arkansas, and begin our exploration of the great overland route to Utah and the Pacific coast.

I will not detain you with the account of our downward trip, but ask you to suppose us, a few hours after our visit at Robinson and Kokomo, snugly "at home" in the station in Poncho Springs, half a dozen miles west of Salida.

XXIII

FROM PONCHO SPRINGS TO VILLA GROVE.

Strength to the weary,
Warmth to the cold,
Blood to the wasted,
Youth to the old;
Ah, and the rapture
Thousand-fold dearer,
Ne'er to be told:
Learn ye the secret,—
Taste ye the sweetness.

THE visitor to Poncho Springs is pretty sure to get into hot water, and, strange to say, the visitor is pretty sure to like it. There are several reasons for this peculiarity, and among the most important is this, that like the wind to the shorn lamb, the water is tempered. It needs to be tempered, indeed, for when one literally gets into hot water, one does not like to have its warmth so emphatic as to make a veal stew of the first leg that is thrust into it. Hot springs whose temperature makes any well-regulated thermometer's blood boil and sends the mercury up to 180° in the shade certainly needs tempering. When properly moderated, however, one cannot fail to enjoy a bath in the soda impregnated waters of the Poncho springs.

The village, to which the springs have given their name, is snugly tucked away in a niche in the Arkansas valley, at the mouth of Poncho pass. The waters of the south fork of the Arkansas river, clear as crystal, and flowing with a foam-flecked current, race rapidly past the town. Along the river's course the cottonwoods crowd, and to their branches, beginning to grow bare, still cling a few trembling yellow leaves. Beyond the river and to the south and west rise the hills, their sides and summits covered with dark phalanxes of pines. Turning one's back upon the town and looking toward the north and west, one sees the snow-crowned summit of the Collegiate range, with all the differences between Princeton and Harvard and Yale entirely eliminated by that distance which ever adds enchantment to the view. Closer at hand, and towering grandly into the sky, a tremendous watch-tower in the west, stands Shavano, while lesser peaks and nameless pinnacles cluster and crowd around. Great plains, broken by buttes, stretch away to the northward, but mountains and foothills circle round to the east and south and west.

APPROACH TO THE BLACK CAÑON.

In this sheltered nook lies the picturesque village of Poncho Springs, and hither do the invalids and tourists flock during the warm half of the year to drink the medicinal waters and to bathe in the healing springs. I strolled through the main street of the town, along which are built substantial frame shops and hotels, and observed evidences of stability upon every side. Poncho Springs is not the result of a temporary craze, nor is it a railway terminus town to be torn down and shipped forward as the road advances. There is a good agricultural country around the village, and the Springs will be a source of permanent prosperity. One of the most picturesque features of this picturesque town is a residence which my traveling companions called "a symphony in logs." The house is to the right of the main street and is built of hewn logs, and with gables filled in with ornamental work, with painted roof and fanciful porticos, presents a peculiarly pleasing appearance.

Passing on through the town toward the hills and crossing the river, one discovers a sign board, upon which it is announced that the distance to the hot springs is three-quarters of a mile. Putting confidence in this announcement, the visitor cheerfully advances along a good wagon road, which soon begins to twine and twist among the hills, at times making a grade of thirty degrees. Finally, just as one begins to lose faith in the guide board, the trail, with an abrupt turn to the right, descends into a gulch, and rises steeply on the other side. Clambering up this steep, the visitor sees to the left the hotel buildings, which announce the presence of the springs, while to the right are pitched a number of tents, late sleeping rooms for an army of summer visitors which, vanquished by cold breezes, has broken camp and fled.

After a bath in the conventional zinc contrivance, to which was admitted hot and cold water through most unpoetic and sternly practical faucets, all of which suggested "modern improvements" rather than a wonderful natural phenomenon, I went out in search of the hot springs, quite as much to re-establish a somewhat shaken faith in their existence as for any other purpose. My doubts soon dissolved, for back of the hotel and half way up the grade of a steep hill, I came upon a little rivulet of soda water still steaming with the heat of its parent spring. A little further on I saw a white tumulus of volcanic formation, and scattered over its summit oval openings in which boiled and bubbled water fresh from Pluto's kitchen. In some of these springs the water was scalding hot, while in others it was merely lukewarm. Springs showing such radical differences of temperature were frequently not more than two feet apart. There are over fifty of these springs here, and no two of them precisely alike.

The Springs have lately passed from their former ownership into the hands of new men, who are very enterprising. Larger buildings have been erected, and the camp-like freedom of the place has been exchanged for something more nearly approaching ordinary hotel life. There is room for about 150 guests, and every requirement for the comfort of invalids. The advertisements issued by the proprietors dwell largely upon the similarity of these waters to those of the Arkansas Hot Springs, and recommend them as equally curative in the special ailments that have long made the Arkansas waters famous.

A few miles northward of Poncho Springs there is a cluster of mining villages, of which the chief are Maysville and Monarch. They lie well under the shadow of the mountains, and silver ore is produced steadily in considerable quantity. These towns communicate with the outside world by a branch line of railway which diverges at this point.

In the quiet of the evening, at this charming retreat—for we had few pleasanter halting places—the Madame bethought herself (seeing the rest of us pen and pencil in hand) that she owed a letter to her Eastern *confidante*, and also remembered that she had promised her an account of our youthful *chef*, with whom by this time we all felt tolerably well acquainted. Happy accident brought this letter under my eye and I seized the opportunity to copy it, so here it is, or at least so much of it as relates to the boy

"My dear Mrs. McAngle:

"I must tell you about our cook; or, as my husband would, no doubt correct me, our *porter*. How our first boy fought and bled and died I wrote you before, and that the last I saw of him he was being bundled rheumatically aboard the homeward train. Well, after I had finished that visit at Pueblo San Juan with old Santiago's wife, whom I described to you, I went home—that is, you know, back to our train—just at evening. As I opened the door a bright-faced boy rose to meet me, with a pair of the most beautiful eyes I have ever seen,—just the kind of orbs young ladies waste oceans of sentiment upon, you know, in boarding-school days. He was, so he told me, a mixture of Kentuckian and Canadian Indian blood. His grandmother, the only one of his family to whom he seemed to feel any allegiance, had set him up in business as a liquor seller. 'But,' he said, 'the business was too rough for *me*, so I gave it up to a friend and came out West.'

"He proved the direct opposite of his predecessor. While Edward could cook, Burt could not; and while Edward had an abhorrence of water, Burt was never so happy as when his pots, pans and kettles were all before him and he was busy scouring. The only difficulty was, that he could not *keep* clean, but was for ever 'clarin' up,' during which process it required considerable ingenuity to make one's way through the débris of the kitchen furniture.

"It was not long before the inside of his car was covered with tinware of all descriptions, pails, smoothing irons, pokers, tools,—everything that could by any possibility be hung up. He had a passion for driving nails, the larger the more fun apparently, for his nails mostly went clear through the car-walls, which soon came to bristle like a newly furnished pin-cushion.

"With an eye to our future interests in all possible contingencies, Burt laid hold of anything along the road that he thought might be of use to us, entirely ignoring any proprietary rights which others might think they had in the object 'smoudged,' as he expressed it. In this way we gradually became possessed of an endless quantity of odds and ends, which it required a decided exercise of authority to get rid of.

"In traveling, he was nearly always to be seen on the top of the car, for he had an appreciative eye for scenery,—so much so, indeed, as sometimes to interfere with his duties. His great fault was procrastination.

"When we reached Durango he became very greatly depressed, and on my inquiring the reason for his melancholy, he attributes it to the dullness of the place; 'for, ma'am, there is no excitement,—no one has been killed for two weeks. Not at all what I was led to expect.' On reaching Leadville, he became much more cheerful, as he had only been in the city six hours before seeing two fights and half of another.

"His gait was something peculiar. It can best be described by the ditty we used to sing, my dear, which commemorates so touchingly the character and adventures of Susanna in her excursions abroad,—

'When she walks she lifts her foot,
And then she puts it down again.'

"Long, lank, dark-skinned, dressed in flapping coat and immensely broad and excessively slouched sombrero until my husband bought him a cap), with his loping walk and swinging elbows, he was easily recognized at a long distance; and as he would come sailing down upon us from afar, with arms full of bundles, he reminded one a little of some huge bird of prey.

"He had a wholesome fear of rattlesnakes and grizzly bears, which the wicked men of the fort maliciously represented to him, abounded in terrible numbers, and of the most ferocious kind wherever we went. 'No, ma'am,' he said, in his slow, stately way, when I cautioned him one day about trying to shoot a bear if he happened to meet one, as they were hard to kill and especially dangerous if wounded, 'No, ma'am; if I meet a bear you just bet I don't stay to take his portrait, but shin up the first tree I come to.'

"He was continually developing new accomplishments. We learned, after a few weeks, that we had not 'prospected' him thoroughly at the beginning. He proved to have had more experience than his youthful looks and aimlessness of motive lead us to expect. We had little occasion to call into use whatever knowledge he had acquired as a bar-keeper, because the education of the gentlemen of the party had not been neglected in that direction,—wholly in an amateur way, and they were accustomed, while 'concocting elaborately commingled potations,' (as they grandiloquently termed mixed drinks) to say to one another: 'If you would have anything well done do it yourself.'

"One day, however, great delight was caused by the discovery that Burt was a barber. His services were at once required, and when, at the end of long labors, he was munificently offered two nickels, he declined them. This noble independence aroused 'Chum's' admiration. He said that he was glad to see that the boy was free from the mercenary spirit so painful to witness in the young.

"Our porter seemed to consider the whole expedition a huge joke, and ourselves a show arranged for his especial benefit. If—as it frequently happened, for a more thoroughly heedless and forgetful youth never existed—we were obliged to expostulate with him on some neglect of duty, a seriousness of countenance would remain with him for some time, but the first joke that came to his ears dispelled it. Sullen, he never was, or ill natured; and if any real emergency occurred, more willing and unselfish help could not have been tendered by a firm friend than was tendered by this servant. I repent me, indeed, Mrs. McAngle, of having made fun of him, even in the privacy of a letter to you. The odor of the steaks that he cooked still lingers in my grateful nostrils; I remember that without him, material for many jokes would have been wanting, and I look on his fast vanishing, but always picturesque figure, with regret."

Standing here at the very foot of the mountains that hid the enchanting netherland of "the Gunnison," we were eager to hurry on to the Pacific slope of the State; but one little side trip remained to be made, and on the second morning we coupled our cars to the express bound for Villa Grove and Bonanza. The course lay up Poncho Pass, and in five minutes the noisy locomotives announced that the ascent had begun.

It was very pretty, as, indeed, we had suspected during our walk the day before up to the hot springs, which stand near its entrance. The track is dug out of the side-hill on the northern side of the gulch, and a bright stream comes tumbling down through willows, cottonwoods, oak shrubs, wild cherry thickets and bushes of service-berry whose crimson fruit tempts you to leap off the train and taste its tart and fragrant juices. The slopes on both sides are covered with evergreens and aspens

"That twinkle to the gusty breeze."

Up through a rift in the trees we catch a glimpse of the little watering-place, and a few miles farther, pass the log-buildings of the old Toll Gate, occupying a pocket in the hills. Only now are the gray carpeted plains of the Arkansas, the village at the mouth of the cañon, and the rough high hills, away beyond the river lost to view. At the head of

Poncho Pass is Mears' Station. It occupies a narrow defile, the walls rising steeply to unseen heights, and the gorges dropping apparently to unfathomable depths. We could not trace the devious course we had come, nor understand how it was possible the railway should surmount the stupendous barrier lying to the westward. Yet we knew that a day or two later our cars would roll steadily to the summit and steadily descend on the other side, for this little nook, the head of Poncho, is only the foot of Marshall Pass, by which the oceanic divide is crossed on the transcontinental route. Nor was it easier to see how we were to get away down the precipitous defiles in which the southern slope of Poncho Pass seemed to lose itself. It was with strongly excited curiosity, then, that we detached from the express and caused our cars to be coupled to the freight train, which the bulletin averred knew how to go down to Villa Grove, and would one day carry the traveller through to Saguache and the South.

When all was ready to make good this promise,—and if that miserably memorable engineer had thrust his shock of hair and bullet-head a trifle further out of the cab-window the company might have dispensed with the headlight—took the back track for a few rods, trended away on a curved side-track to the right as far as the hillside would admit, crossed the main line on a bridge, and having by this time accomplished a half circle, headed eastward again and began to climb the southern side of the gulch in a line so parallel with the lower track that a mile later you could fling down a stone from one to the other though you were a couple of hundred feet above. Half a mile more and the summit is reached,—a green saddle between the foothills of Mount Ouray on one side and the far-braced buttresses of Hunt's peak on the other. The going down is fairly straight and easy work, and it is not long before the gulches widen out, the diminished, grassy hills are left behind, and your speed increases as you strike the firmly bedded, regular track, pointing southward through a broad, treeless plain.

Perhaps I have said enough of the wonderful beauty of the Sangre de Cristo range, seen from this side; have too often told of their compact array and unbroken grandeur; of the scores of nameless peaks that vie with Hunt's, Rito Alto, Electric, the gothic Crestones and the group of pinnacled, sun-gilded summits that crowd near far-away Blanca; but in the broad morning light of this day's trip they stood up in freshened color and renewed majesty. All the cloud-curtains were rolled up, and heaven shed unhindered its clear, sharp sunbeams from end to end of the magnificent chain. The souvenirs of yesterday's storm added decoration, for the summits were all dusted and powdered, with light snow, like noble heads of the old *regime;* and this unwonted covering descended far enough below timber-line to frost the upper lines of trees, so that there was a soft gradation from the deep verdancy of the lower slopes, through hoary greenish-gray to the unbroken white of the clear-

THE GUNNISON.

cut gables lifted into the serene and absolute solitude of the cærulean dome.

Down between the sharp-edged spurs come numerous streams, watering little spaces where ranchmen had placed their cabins and fenced in their fields. Large areas now given up to the badgers and sage-brush, can be brought under irrigation, when the more favorable lower parts of the valley have been utilized. A broad road runs down here,—the old wagon road from Leadville and the Arkansas valley to Saguache, Del Norte, the San Juan region and New Mexico.

The same words apply to the more broken western side of the valley, here called Homan's Park, though it is only the upper end of the San Luis valley; and, in addition, those western hills are full of prospectors, and of places where prospecting for silver and gold has met with success. This is the celebrated Kerber Creek district, and Bonanza, Exchequer, Sedgwick and other little centers of human interest, lie back of those rugged, green hills over which the angular heads of Exchequer and Ouray mountains stand in high-chieftainship.

Of all these, Bonanza is the largest. The ores, however, are characterized by being of a low grade, but great volume, and by containing refractory elements, with a small percentage of lead, so that the large smelter at Bonanza has been compelled to cease running until it could provide itself with a more adequate outfit of fluxes, etc.

Villa Grove—a pleasant little village on San Luis creek, which drains the upper part of the park—is the railway point for all these mines and several other settlements not yet mentioned. Looking southeast from the station you can see where the track runs up into the foothills of the Sangre de Cristo to one of the great iron mines of the Colorado Coal and Iron company, whence large shipments are being made daily. Though of great importance and value, the seeing of this mine amounts to little, since it is hardly more than an open quarry.

From Villa Grove stages leave daily for Bonanza, Saguache and half a dozen other places, such as Crestone and Oriental,—little mining camps in the foothills. The roads are so smooth and level everywhere that the great six-horse Concords are unnecessary and spring wagons are used.

XXIV

THROUGH MARSHALL PASS.

> Then felt I like some watcher of the skies
> When a new planet swims into his ken;
> Or like stout Cortes, when with eagle eyes
> He stared at the Pacific—and all his men
> Looked at each other with a wild surmise—
> Silent upon a peak in Darien. —KEATS.

NE of the wonders of Colorado progress is the Gunnison valley. The "Gunnison," as it is usually termed, embraces a wide area, being, in popular parlance, everything in Colorado west of the Continental Divide, north of the San Juan mountains and south of the Eagle River district. In fact, this is correct enough, for nearly all this great region is tributary in its drainage to the Gunnison river,—the third great stream which unites with the Grand and the Green to form the Rio Colorado. The water-shed between it and the Rio San Juan, the only other feeder of the Rio Colorado worthy of mention, is the very high and wintry ridge of the San Juan mountains, crossing which you find yourself in Baker's Park and the region we had just come from. Betwixt the head of its northern branches and the springs that feed the Grand River basin, stand the Elk mountains and the high table lands of the Grand Mesa. From the one water-shed to the other it is about fifty miles.

Ten years ago this region had hardly a wanderer in it from one season's end to the other, and was full of Ute Indians. There were two or three agencies, and roads leading thereto, but it was all a reservation. Everything civilized that entered the district came up from Saguache through Cochetopa Pass and along Cochetopa creek into the Uncompahgre valley, where the Utes spent their winters. There was also a trail, occasionally traveled by sportsmen and explorers, leading southward from the Los Pinos agency to the headwaters of the Rio Grande and on over Cunningham Pass into Baker's Park. I marched over it in 1874, and a cruel march it was, though full of picturesque interest. An Indian trail northward to White river was about the only other internal pathway. The region, therefore, was a *terra incognita* to Coloradoans, as well as to the rest of the civilized world.

But this mystery was soon to be cleared away. The search for gold and silver, which has led to more exploration of unknown regions than all the geographical societies of the world put together, did not hesitate to encounter the darkness that overspread the Pacific slope of the State, and to go prospecting thither as soon as ever a hope of finding "mineral" entered into the miner's heart. Close following upon the rush to Leadville, was repeated the history of the Pike's Peak sequel. Now, as then, men disappointed in not finding mines of fabulous wealth, during the first week of their stay, or shrewdly thinking to anticipate the crowd, began to walk further and further afield in search of new argentiferous rocks, so that by the summer of 1879 we began to hear not only of Ten Mile and Red Cliff, but of the Gunnison, as a district where success had met the prospector. That was only a little over four years ago. Now how well are we acquainted with this erst mysterious and Indian-haunted valley! Four years ago a mule was the best mode of conveyance hither, and an Indian trail almost the only pathway. Yesterday I rode into the heart of it in a parlor car, and found, ready for my perusal, the morning newspaper, with a day's history of all the world, from Chicago to Cathay.

The Gunnison country boasts several towns of considerable size, some of them the center of a circle of mines which radiates from them, and from which they absorb cash and conviviality. First in size is Gunnison City; and after it in importance are Crested Butte, Lake City, Ouray, Montrose, Delta and Grand Junction,—the last three being situated in the old Ute reservation in western Gunnison. Of less size, but yet centers of population, are a large number of small mining towns or "camps," such as Ruby, Crooksville, White Pine, Pitkin, Irwin, Barnum and Ohio. Each of these would require some attention from a faithful chronicler of the county, for they are all in Gunnison, where the territory is large enough to enable one to set in it the whole State of Massachusetts without crowding—that is if you lopped off Cape Cod or curled it up into Marshall Pass.

It is by the way of Marshall Pass that the railway enters the Gunnison. Leaving the main line and the Arkansas valley at Salida, only five miles are traversed before the train begins to enter Poncho Pass and climb the mountains, which it requires four hours, express speed, to cross,—four hours of uninterrupted pleasure.

Of Poncho's prettiness I have already spoken. Its summit is found at Mears' Station, and then begins the real ascent of the Continental Divide. In a few moments the circling rim of the pit-like valley is surmounted, and Hunt's peak, by its cap of snow signifying its superiority to the giants of the Sangre de Cristo about it, rises like a planet over the hills we are leaving behind. We seat ourselves on the rear platform and watch it until the whole range, of which it is the northernmost officer, stands drawn up in purple line before us, and we can trace the summits, pressed back into straight line, for perhaps sixty miles to the southward—

"————Sierras long,
In archipelagoes of mountain sky."

What can that goodly rank, each peak sharp and pyramidal just along side of the other, every curve of the foothills parallel with the one before, sweeping down into the trough-like park at their hither base,— what can all this uniformity be but the splendid chain of the Sangre de Cristo? Have we not seen it time and time again, beheld it from east and west and south, and now here from the north; and has it ever been out of line, or anything but a soldierly array of uniform heights bearing the same relation to the ill-assorted army of the rest of the Rockies, that the famous grenadiers of Frederick the Great did to his peasant conscripts?

This sight explains to us also, that the great width of lofty hills we are picking our way through now is the junction mass of two ranges. It is here that the Sangre de Cristo starts off on its own line to the southeastward, while the main chain, forming the backbone of the continent, trends somewhat westward and continues to do so more and more till it loses itself in the jumble of San Juan, San Miguel, Uncompahgre, Bear and other ranges that fill the southwestern corner of the State. The summits north and southwest of us divide the Atlantic from the Pacific; but that magnificent corps that will not be left behind, but seems to march steadily after us, in battle array, separates only the Arkansas from the Rio Grande. The glimpses of valley we get now and then just this side its base are of Homan's Park, which is only the upper end of the wide San Luis, and places can be seen that we could not reach by long traveling.

Marshall Pass itself, which we enter imperceptibly out of Poncho, is a depression in the main range and lies between Ouray and Exchequer mountains. It was a daring scheme to run the road over here—for *through* wouldn't express it properly. The summit is almost eleven thousand feet above the sea, and timber-line is so close that you can think sometimes you are actually there. The trees are stunted and all stand bent at an angle, showing the direction of the fierce and prevalent winds that have pressed upon them since their seedling days. The cones they bear start bravely, but after perfecting three or four broad circles of scales and seeds the nipping frosts of August and September admonish them to make haste; so the remainder of the cone is put forth so hastily, in Nature's attempt to complete her work, that the whole remaining length of fifteen or twenty circlets will not exceed the length of the first two or three full-grown scales, and the cone ends ridiculously in a little useless acuminate tip.

To attain this height, the road has to twist and wriggle in the most confusing way, going three or four miles, sometimes, to make fifty rods;

but all the time it gains ground upward, over some startling bridges, along the crest of huge fillings, through miniature cañons blasted out of rock or shoveled through gravel, and always up slopes whose steepness it needs no practiced eye to appreciate. To say that the road crosses a pass in the Rocky mountains 10,820 feet in height is enough to astonish the conservative engineers who have never seen this audacious line; but you can magnify their amazement when you tell them that some of the grades are 220 feet to the mile.

The mountains and hills in the neighborhood of Marshall Pass are clothed for the most part with grass, or else sage-brush and weeds, and with timber, scant in some places, dense in others. The tourist will not see there the startling cliffs and chasms that break up the mountains on the road to Durango, but, on the other hand, he will not feel any terror at dizzy precipices, nor tremble lest some toppling pinnacle should fall upon his fragile car. No better exhibition of the greatness and breadth of these mountains could be found, however, than here. There stretches away beneath and around you an endless series of hills, some rounded and entirely over-grown with dark woods, others rising into a comb-like crest, or rearing a dome-shaped head above the possibilities of timber-growth and covered with a smooth cap of yellowish verdure. They crowd one another on every side, and brace themselves, each by each, as though their broad and solid foundations were not enough for safety. They stand cheek by jowl in sturdy companionship, taking rain and sunny weather, hurtling storms and serene days with impartial equality. Your vision will not find the limit of these huge hills until it is cut off by the serrated horizon of the crest of the Sangre de Cristo, or by some frowning monarch near at hand, holding his head high and venerably gray, as becomes a chieftain, where he can get the first messages of the gods and be looked up to by a thousand of his more humble kin.

"It is like a huge green sea," murmurs the Madame, hitherto silent with gazing. "I know a great many people have made the same comparison before—have often said that these commingled ranges were as a sea, tossing its white crests here and there and all at once congealed; but that is the very impression which fixes itself upon you. These rounded, or sharp-edged, tumultuous mountains are like a wide, green ocean." The great cone on the northern side of the track, close to which the roadway skirts nearly the whole distance through the pass, is Ouray peak. Ouray, as nearly everybody must know, was the head chief of the Utes. This tribe only very lately abandoned all this portion of Colorado, leaving last that reservation which lies beyond Gunnison City, and which we are soon to visit. The peak we have hugged so closely does honor to the dead chief. The farther you get around it the more nobly do its proportions rise into the blue ether. Like Veta mountain, which it closely resembles, this peak is of white volcanic rock that has decomposed into small blocks. The sides then are loose

CURRECANTI NEEDLE, BLACK CAÑON.

"slides," as steep as the fragmentary stuff will lie, and the top is a narrow summit with smooth, rounded outlines. We are only a few hundred feet from the topmost timber, yet the bald white summit rears its head to almost unmeasured heights above, and claims our admiration by its simple majesty, far more than does the broken, cliff furnished upthrust of Exchequer peak opposite, though its black head is held quite as high. Perhaps this is only because we have become somewhat tired of the closely-shutting high mountains; weary of being

" ——under ebon shades and low-browed rocks,"

as Milton puts it. Certes, it *is* good once more to be able to look abroad!

On our way to the summit we had crawled through long snow-sheds, built to protect the road from the snows of winter, and which are hung late in spring with brilliant icicles formed by the sun without and the cold within. Passing through the last shed, which has a length of fully half a mile, we reached the highest point of the divide, and while the extra engine which had helped pull us up the steep grades went cautiously down the valley toward Gunnison before us, we climbed the rocks about the little station house, to enjoy at its best the magnificent view presented. To the northeast, white with snow, towered the serrated range of the Sangre de Cristo mountains, rising abruptly from the valley which stretched away to the southeast and standing out in bold relief against the deep blue sky. Between the range and us were lower hills and isolated peaks tumbled into a confused mass, and only prevented from pressing too closely together by the little valleys that ran between them. Immediately around us grew stunted pines, bent, barren, blackened and lifeless. Down the mountain side the forests became denser, greener and fresher, while from the distant valleys, at the bottom of which we could see tiny streams working their way to worlds beyond, came low murmurs and sweet odors. Toward the west, and losing itself in a hazy distance, ran the Tomichi valley, narrow, heavily wooded, and free from all that rocky harshness so prevalent in Colorado. Far below we could look down upon four lines of our road, terrace below terrace, the last so far down the mountain as to be quite indistinct to the view. The iron loops were lost to sight at times as the road wound about some interfering hill; and often the forest was so dense that the track seemed to have disappeared for ever. Five hundred feet down the mountain side we could see a water-tank, and knew that it marked the spot where we would be, after an hour of twisting down the incline. As we gazed upon the mountains, the valley, and the far and farther heights, we could imagine ourselves returned to the beginning of things, and shown the globe only that moment finished. There was a wealth of coloring, a sublimity unsurpassed, and withal an attention given to detail by which the picture was made perfect. I remember to have stood on Marshall Pass once when the sun was just dropping out of sight beyond the rolling hills to the westward. As it sunk lower and lower behind its curtain of snowy peaks, prismatic hues came flashing along the pathway of its fading light, which touched the rugged sides of Ouray peak and the white-capped range beyond until every treeless spot and gabled peak shone with a mellow hue. All objects—those near by and those far away—flashed bright colors, beautiful, brilliant, and as varied as those of the rainbow. From the mountains long shadows were cast, and in the forest crept dark shades. All nature prepared to sleep, and no sounds came from around the lonely pass but the sighing of the wind as it swept through the tangled trees. "All outward things and inward thoughts teemed with assurances of immortality."

Our descent from the pass was continuous but slow. At least it was slow at first. All steam was shut off in the engine and the air-brakes were used to preserve a uniform speed. Winding in and out among the trees, and catching at different times extended views of the Tomichi, we worked our way to more level country and were soon skirting the meadows and whirling across the ranch properties of the fertile valley. Close beside us ran a sparkling stream, tapped here and there by the farmers, who used its water for their lands, and again winding its way through the willows that grew on its banks. Looking back over the way we had come, there appeared dark-green forests, backed by high mountains with bared summits; but before us lay the Tomichi, shut in on either side by low hills and extending westward so far that its end was lost in haze. Everything was green, fertile, luxuriant. Cattle grazed in the meadows, ricks of hay stood by the side of low-roofed cabins, and narrow valleys came down from the northern mountains to join the one along which we kept the swift and even tenor of our way.

XXV

GUNNISON AND CRESTED BUTTE.

> "Over the Mountains of the Moon,
> Down the Valley of the Shadow,
> Ride, boldly ride,"
> The shade replied,
> "If you seek for El Dorado."
>
> —Edgar A. Poe.

AT its lower end, as the mountains in the range we have crossed begin to grow indistinct in the distance, the Tomichi valley pushes aside the hills which have hitherto confined it, and broadens into a wide, grassy plateau, encircled by mountains, in the center of which stands Gunnison, the chief town of Western Colorado. Westward, where the river comes down, sculptured cliffs rise near and abrupt; but elsewhere the mountains are far away enough to make invisible all their lesser characteristics. Those to the north and south east have their long line of irregular summits capped with snow; but to the west the ranges grow less rugged and more rounded, while between the hills runs the valley occupied by the Gunnison river on its way to the Grand, and by which the railway enters the rich farming lands of the newly opened reservation and the territory of Utah.

Drawing rapidly nearer the center of the plateau, we approached the city and perceived that it consisted of two distinct parts, with a gap of half a mile between them. Then a new freight-house cut off the view and we came to a stoppage in one of the busiest "yards" outside of Denver.

The town, as I have said, stands in the middle of a level park, at an altitude of about 8,000 feet above the sea. There is room enough "to hold New York City," as the people are fond of saying. No stream waters the middle of this area, but skirting the further edge, just under the bluffs, which on every one of these bright summer evenings

> "—— topple round the West,
> A looming bastion fringed with fire,"

runs the Gunnison river, through a bosky avenue of full-foliaged trees and thickly interlaced underbrush. Away to the southward of the town again, the Tomichi curves about the base of rounded, plush-tinted hills that look like the backs of gigantic elephants. I have called the first of these streams the Gunnison, but if you follow it up a little way you will come to repeated forkings known as East river, Taylor river, Ohio creek, and so on. I believe, therefore, that properly the Gunnison does not attain individuality and deserve its name until all this cluster of northern tributaries joins with the Tomichi, just below the town, and the united and largely increased stream flows independently onward.

A UTE COUNCIL FIRE.

It is in the fork of these chief sources of the Gunnison,—at its very head so to speak,—that the town is placed. It is not upon the banks of either, but the pure waters are easily led in open aqueducts all over the site, running by their own current. There are places enough for them to run, too, and people enough to consume them, leaving only begrimed tailings for the engine-tanks at the station.

The town began in two parts and became the shape of a dumb-bell, the handle represented by Tomichi avenues. The knobs of town at each end are rival districts known as Gunnison and West Gunnison, but the former is the larger, seems to have the start, and has secured such distinctions as the post-office, the banks, the court-house, the high school and the principal newspapers. These, with several of the mercantile establishments, show fine structures of brick and stone, the latter being a white sandstone of great excellence for building purposes, which abounds in the buttes on the edge of the plateau. The majority of the houses, both for business and for residence, however, are frame buildings. Some are of pretentious size, and many prettily decorated, so that we do not know, a cleaner, more regular, cosy-looking city in the state than this. The divided appearance is gradually disappearing by increased building between, which proceeds with amazing rapidity.

The history of this valley and town is entertaining. In the early days of Rocky Mountain exploration, this whole region was known as the Grand River country, its noblest stream, now called the Gunnison river, then being known as the South Fork of the Grand. Of its history, or its geography, as I have intimated, little has been known until very recent times. It is recorded that in 1845, ex-Governor Gilpin, then a mere lad, traversed the entire length of the river-valley from west to east, on his return from Oregon to St. Louis. "Having crossed southern Utah by an old Spanish trail he pushed his way up through the valleys of the South Fork of the Grand, crossing the divide very near the southeastern corner of what is now Gunnison county. Although pursued relentlessly by savages he was enthusiastic over the results of his trip and embodied the knowledge so obtained in a map which is now on file in Denver. The interval following Governor Gilpin's exploration between 1845 and 1853 is entirely an historical blank, only vague Indian stories being given out by occasional trappers and by the Mormons, who joined in relating the beauty and richness of the country.

"In 1853 Captain Gunnison, a gallant officer, following Rock creek up to its head, discovered a nobler stream coursing southward from the Elk mountains. This stream cost him his life. As he was exploring it, he was set upon (whether by Indians or not seems doubtful) and cruelly murdered. After this adventurous officer the Gunnison river was named and afterwards Gunnison county. In 1854 the indomitable 'Old Pathfinder,' General Fremont, passed over nearly the same country from east to west and in his report paid glowing tribute to the beauty and wealth of these regions. It was not, however, until 1861, when some prospectors, approaching through California gulch, where Leadville now stands, gave names to Washington gulch, Taylor park, Rentz's gulch, and Union park, that any positive development was undertaken. Then it was only on a very small scale, and although the discoveries they made created considerable excitement in mining circles, the fear of Indians was yet so great as to prevent any immigration of any consequence. This fear was heightened by the horrible discovery one morning of the massacre of twelve men in Washington gulch. This wholesale tragedy gave a gloomy side-defile the name of Dead Man's gulch. The story of this outrage quickly spread throughout the entire country, each person coloring it as it went and adding a little to the horrors of the event. At this time nothing could tempt the daring miners of the adjacent and already populous Colorado gulches to risk their lives in this country. Even the most marvelous stories which were told of the golden bullets used by the Indians, and of mines to which El Dorado and Comstock and Golconda were vanities, failed to tempt their cupidity sufficiently to cause them to venture into the blood-christened country. A few, however, who had already forced their way in, earned a precarious livelihood in Washington gulch, fortified from the Indians and living for months at a time upon game and fish. In their leisure moments, between fighting Indians and hunting game, they occupied themselves in placer mining and, it is said, made from five to twenty dollars a day. Not until 1872, however, was any organized attempt made to open up the country. In that year Jim Brennen, of Denver, headed a small party of prospectors and located in the Rock Creek region.

"From this time really dates the origin of the mines, their reports being so enthusiastic that in 1873 Dr. John Parsons, Professor Richardson and thirty miners entered from Denver. One of the stories of this party which is told, but which is historically doubtful, runs to the effect that in pushing around by the southern entrance over the Saguache, General Charles Adams, who was then in charge of the frontier, forbade their further progress without the consent of the Utes. A heated debate is supposed to have arisen over the matter, which was settled by Chief Ouray himself, voting to grant them permission. In 1874 a colony was formed in Denver to settle upon and cultivate the Gunnison's agricultural lands. Accordingly twenty men, all told, located themselves at various points upon Tomichi river and gave their special attention to ranches. The mining districts, however, on account of the Leadville and San Juan excitements, together with the difficulties and inconveniences of mining in this country at that time, did not really begin to grow until several years later."

In the latter part of 1877 the state legislature set off Gunnison county, containing about twelve thousand square miles, or an area somewhat larger than the state of Connecticut. Three-fourths of it lay within the Ute Reservation, and it has since been subdivided into four new counties,—Gunnison (restricted to the eastern end), Montrose, Delta and Mesa. By 1880 matters began to assume a fixed condition. The people left their tents and sought more durable habitations. Business ceased to be desultory. The prospect-diggings, of which five thousand had been recorded, were developed as rapidly as possible, the buzz of the saw-mill and planer was heard, and smelters began to be erected.

Historically, there is little to add. Steady growth has benefited the city. New and large business blocks have been erected, a handsome hotel built, and a smelter put in operation. It has now a population of fully five thousand, is lighted with gas, and has a system of waterworks. The streets are wide and clean; and the entire town has lost that frontier appearance which characterized it in its earlier days.

And Gunnison is a railway center. To the north the Denver and Rio Grande has extended a branch to Crested Butte and brought into closer communication with the outside world the adjoining mining towns of Irwin, Ruby, Gothic, and others of less importance. The road leads northward from Gunnison up the pointed valley until it gets close upon the bank of East river. Following the river, the valley narrows into a ravine, and some interesting masses of broken volcanic rocks, injected edgewise into the general sandstone strata, attract the eye.

It is the far-away landscape, nevertheless, that holds attention as we look backward. Rising above the level of the plain upon which the city is built, you can span with your vision hills and mesas southward, and behold "striking up the azure" a vast length of the ever-magnificent San Juan mountains,—the same glorious pinnacles that towered about us, near at hand, in Baker's park. We could count the peaks by dozens if we tried, but it would be rash to try to name the separate points of the long serration. Many snow clouds have shed their burdens upon them since we saw them last, but to-day their heavens are clear and the sun blazes down upon scores of miles of lofty névé fields, the uniform purity of which, at this distance, seems broken only by the shadows the higher peaks throw upon their lowlier companions and upon their own half-concealed sides. Gazing at them across the dim foreground of sage-plain, the middle scene of receding, intermingled, haze-obscured and bluish hills, we were more and more delighted with their loveliness,—a word whose propriety you will appreciate when you, too, have laid away this treasure of memory—one of the most entrancing bits of landscape in Colorado.

There are a few patches of rank meadow, but most of the way the hills run down so close to the river banks that there is barely room for the road-bed to be made. Growing so close to the water that they are reflected in its depths, are sweet-smelling trees, tall, graceful, luxuriant, but in winter they bend beneath the snow that clings to them. Reaching to the top of the hills and completely covering them, are tangled masses of brush, pushed aside at times by forests of pines and torn asunder in places by the rocks that have lost their balance on some far summit and been rolled to the river below. In the narrowest places precipices menace each other across the stream; and on their faces, brown and weather beaten, grow hardy shrubs, clinging to the crevices and hugging the bold headlands.

Nor does the valley afford satisfaction to the lover of what is only picturesque in nature. We have seen many a trout whipped from his cool retreat under the shadow of the rocks. The region is a sportsman's paradise. Nature is at her best, the forests are full of health-giving odors, and a day's tramp could not fail to bring color to the palest cheek, strength to the weakest body.

Twenty-eight miles north of Gunnison the narrow valley lets us

OURAY.

into a snug little basin among the hills which border upon the Elk range, Slate river comes winding through it from the north, while Coal creek sweeps abruptly around a lofty spur at the left. Straight ahead, behind a green ridge, a white conical mountain stands challenging our admiration, and on our right a still nearer height rises like a mighty pyramid of gray stone from a richly verdant base.

The Madame gazes at them with delight a moment, but quickly glances with more eager interest to the meadow-land in which we are coming to a standstill, for the lush grass is dyed with innumerable flowers.

"Why Crested Butte?" she asks as the station-sign comes in view.

I point, for reply, to the conical gray height which dominates the valley.

"That is neither a butte, nor is it crested," she says. "A butte properly is not a peak of volcanic or primitive rock even if it is isolated —the proper name for that is 'mountain' or 'spur.' A butte is a hill of sedimentary rock, not mountain-like in appearance, and standing by itself in a flat region. Moreover there isn't a bit of crest. Its apex is as sharp and round as a well-whittled pencil."

"If you could look at it from the other side you might find a very well-marked crest."

"But I can't, and nobody does, see it from the other side. However"—and here her prerogative of inconsistency was exercised—"I am glad they adopted the mistake for now the town has a name worth remembering, something you can't say of too many of these mountain villages."

Crested Butte had the honor to be the first settlement in the Gunnison region. A recent review of its history says that in the spring of 1877 the Jennings brothers, who were hardy prospectors, penetrated as far as the Butte and were somewhat surprised and delighted at finding coal. Instantly turning their attention to that branch of mining they located some land. The fame of this discovery, blending with that of others, proved an incentive to the overflow from Leadville and the rest of Colorado. In 1877 a few men came in, but no effort was made even to survey the country until 1878. In that year Howard F. Smith dropped in and purchased some coal interests. He soon had the country surveyed, erected a store and advertised so well that within a few weeks a village had been started which is now one of the pleasantest summer places on the western slope, and can boast a hotel that has no superior in the Rocky mountains for comfort. This is the Elk Mountain house, and it is the property of the town-site company, who appreciate that the first impressions of a traveler (and possible settler) are largely colored by his early experiences in the matter of food and lodging.

No mines for gold and silver exist in the immediate vicinity of the

SOME COAL STATISTICS.

village, though many "camps" in the Elk Mountains from five to twenty miles away are tributary to it; and the chief reliance and *raison d' être* of the settlement is found in the coal-beds that are adjacent to it. These are of the greatest value and importance, and at night, when the blaze of the coke ovens sheds a lurid glare upon the overhanging woodlands and the snug town, one can appreciate the far-seeing expectations that lead the people there to call their town the Pittsburgh of the West.

Between two great foothills south and west of the town, flows a little creek whose channel is cut through five beds of coal, dipping southward, with the rest of the stratified rocks, at an angle of about six degrees; the lowest is ten feet in thickness, the others six, five, four and three feet. This coal is bituminous, and has been proved to be the best coking coal in the United States, as is shown by the following authoritative analysis.

Coal......	.44	......	34.17	......	72.30	......	3.09
Coke......	1.35	......	—	......	92.03	......	6.62

The railway having reached Crested Butte, the coking veins are now well opened "by three drifts on water level, working the seam to the rise." The mines are prepared for an output of four hundred to five hundred tons of coal per day, and the coke can be furnished to any extent, by the Colorado Coal and Iron Company, who own the mines. At first this coke was made in open pits, but now a long series of ovens has been built, and the railway tracks run to the ovens and almost to the mine-entrance. The cars drawn up the incline from the "breasts" to the surface, are thence dragged by mules through a quarter of a mile of sheds, built to guard against the deep winter snows, down to the ovens and the cars. Forty or fifty miners are employed at present, and these live with their families in large log houses built under the edge of the forested hill close to the mine.

The coke of all coal, being composed of fixed carbon and ash, depends for its value on the minimum of ash. The coke from the coal of Crested Butte contains from two to six per cent. less ash than the coke of the best eastern coals, its total of ash amounting only to six per cent. For all purposes of steam, this bituminous coal is said to have no superior on the continent. A well known mineralogist is reported to have said of it that, while a pound of Pennsylvania anthracite will make twenty-five pounds of steam, a pound of this bituminous coal will make twenty-three pounds; but while one pound of eastern anthracite is burning, two pounds of this will burn. Therefore, while the pound of Pennsylvania anthracite is making twenty-five pounds of steam, this coal will generate forty-six.

These coal-beds can be traced without difficulty up Slate river, exposed here and there in the western bluff, and can be found hidden in the opposite hills. As it is followed, however (rising in altitude with the upheaval toward the mountain-center), a change is seen to take place in its character. Two miles above the village it is neither soft nor hard; a little farther on, a part of the bed is decidedly anthracitic; while four miles above Coal creek, and at an elevation of a thousand feet or so above it, genuine anthracite of the best quality is mined from the same seams that, four miles below, yield the coking soft coal. "Nothing could be plainer, nor more beautiful to see," than this practical demonstration of how under different conditions of heat and pressure, the same carbonaceous deposit becomes bituminous or anthracitic.

The anthracite mine is at the top of a wooded hill and is reached by one of the most entertaining roads in all Colorado. The coal beds form strata right across the hill, so that the miners can run their tunnels out to daylight in any direction, and need not fear the gas which is so troublesome in the bituminous diggings below. The vein now worked is five feet thick at the entry, but increases to ten feet in thickness within. It is solid and pure, and is thrown down by blasting. The men are paid seventy-five cents per ton for breaking it into convenient pieces and loading it into the little cars. These cars are then drawn to the brow of the hill and dumped into larger cars which travel on a tramway sixteen hundred feet long, and most skillfully erected on a curved trestle, down to the breaker at the river level. This breaker is the only one west of Pennsylvania, and is capable of transmitting five hundred tons a day, properly crushed, to the railway cars, which run underneath its shutes.

The highest excellence is claimed for this anthracite coal by its owners, not only for domestic purposes, but in the making of steam. In price, this company is able to meet the Pennsylvanians at markets on the Missouri river, and to furnish all nearer points at a much lower rate than eastern shippers can afford; while they hope to secure a large part, if not the whole of the California business, which amounts to about fifty thousand tons annually. The mine and breaker have now been put in shape to yield steadily a large product; they are hereafter expected to be able to meet the whole demand. The anthracite beds in this region are believed to be very extensive, so that undoubtedly other mines will be opened as soon as a large enough demand will justify it. The discovery of these anthracite beds caused an immense excitement, for it was the first true hard coal found in the State; and a mob of men rushed in as though to an old-fashioned placer-find.

This region in 1879, indeed, caused a great flurry in the minds of prospectors who began to enter it at the risk of their lives long before it ceased to be an Indian reservation. As long ago as 1872 argentiferous quartz had been found in Rock creek just over the divide between these waters and the Roaring Fork of the Grand river, where Galena, Crystal, Treasure and Whopper mountains are seamed with large veins of comparatively low-grade, but easily smelted galena ore. The center of this district is Crystal City, and from that point prospectors pushed their way right and left as fast as they dared, and thus led to the opening of the Gunnison region.

It was not until 1879, however, that the precious metals were found in the southern slopes of the Elk mountains, and the region in which we are now interested was heralded abroad as the long-awaited El Dorado. Hundreds of men flocked in, striving to be first on the ground. A few of the earliest comers chose a spot at the base of the sharp, white mountain so plainly in view north of Crested Butte, and decided that a town must be placed there to be called Gothic—a name suggested by the appearance of some cliffs near by. It was done, and the people came to fill it. To it came all the business of the Brush creek, Rock creek, Copper creek, Sheep mountain, and Treasure mountain silver and gold mines, besides those nearer at hand—Schofield, Galena, Elko, Bellevue and others.

Another somewhat separate mining locality was one that we looked down upon as we stood at the mouth of the anthracite tunnel and gazed across the deep gorge which sank between this and the opposite hills, and down which flowed the gentle current of Slate river. The wall on the other side rose above the line of timber growth, and one peak showed an exposed face of brilliant red rock in high contrast to the blue-gray of the rest. Beneath it lay Redwell basin. At the right frightfully rough cliffs and forested crags shut in Oh-be joyful gulch, at the head of which, just out of sight, was Poverty gulch, while Peeler basin showed its edge. It seems to us that we can perceive through the clear atmosphere every tree and stone and crevice on the opposite slopes, though miles away, and can almost hear the prattle of the great waterfall that shines white in the shady bottom of the gorge; but we can see no signs whatever that a human being has ever been in all that area. Nevertheless over all that mountain side there is said to be scarcely an acre of ground not partially covered by mining claims, and upon some part of each one of these a discovery-shaft has been sunk. Many of the fissures thus disclosed are of immense size, carrying veins of argentiferous galena from three to nine feet in width, assaying on the surface from forty to one hundred and sixty ounces of silver to the ton. In some cases ruby silver or gray copper have been reached at forty or fifty feet in depth, assaying over one thousand ounces. At night the coal men see the opposite mountains dotted with camp-fires, and the merchants of Crested Butte will tell you that many a wagon-load and train of burros is packed with provisions for those apparent solitudes.

"What's in a name!" exclaims the Madame as we are riding homeward, while talking over these districts and discussing the notable properties.

"Generally nothing," it is replied, "so far as the designations of mines are concerned, but from the prevalent style of names in the whole district it would be possible to judge something of the men who settled it. Here, for instance, one can't help noticing an absence of the rough gambling titles so common among California mines. The 'Euchre Decks,' the 'Faro Banks,' the 'Little Brown Jugs,' etc., are few, and in their place we find the 'Shakespeare,' the 'Iron Duke,' 'Baron De Kalb,' 'Catapult,' and others with similar literary, historical or mythological meanings. It is evident that no rude typical miner presided at their christening, but that intelligent, and in many cases highly educated, men discovered and named them.

Eight miles northwest of **Crested Butte** are the almost united towns of Ruby and Irwin, which, **in 1879 and '80, had** "booms," but now are **almost** deserted. The neighborhood abounds in silver, but it has been **found that too** many obstacles stand **in the way** of successfully working **the mines, which are** very high, and in a region famous for its deep **snows, until** the science of ore-treatment has progressed, and cheaper methods of operation and transportation have been devised.

Leaving Chum to take the Madame and the train back to Gunnison, I left **Crested Butte on the** morning after our ride to the anthracite mine, **on my way to Lake City,** discouraging all company.

GATE OF LODORE.

XXVI

A TRIP TO LAKE CITY.

*A wild and broken landscape, spiked with firs,
Roughening the bleak horizon's northern edge.*

—WHITTIER.

LAKE CITY is a mining town at the foot of the San Juan mountains thirty miles south of the railway station of Sapinero (the latter named after a sub chief among the Utes who was looked upon by the whites as a man of unusual sagacity). It was at that time reached by a buckboard, carrying the mail and passengers.

The stage-road led up a long, long hill to the top of the mesa between the Cochetopa and the Lake Fork of the Gunnison. This much of the way was in the track of the old southern road to California, which came up from Santa Fe to Taos, San Luis park, Saguache, and so on over here along the Cochetopa, striking the Gunnison river just above this point and continuing on down to the Uncompahgre, where it crossed the Gunnison to the northern bank and pushed westward to Utah. This was the route followed by Captain Gunnison in 1853, and it came to be known as the Salt Lake Wagon Road; and the whole course of the Denver and Rio Grande railway follows it closely from Grand Junction to the Wasatch mountains. The road is still occasionally traveled for short distances by light wagons and by men driving bands of horses, who wish to escape paying the tolls demanded along the new and improved roads, so that it is in no danger of becoming obliterated.

From the top of this high plateau, a great picture opens before the eye, in all directions. Northward the peaks of the Elk range form a long line of well-separated summits. Northeastward, the vista between nearer hills is filled with the clustered heights of the Continental Divide in the neighborhood of the Mount of the Holy Cross. Just below them confused elevations show where Marshall pass carries its lofty avenue, and to the southward of that stretches the splendid, snow-trimmed array of the Sangre de Cristo. They fill beautifully the far eastern horizon, and end southward in the massive buttresses of Sierra Blanca, of which no more impressive view can be had than this elevated standpoint affords. As we advance a few miles other mountains rise into sight straight ahead,—that is, in the southward. These are the cold and broken summits of the Sierra San Juan; while isolated from them, and a little to the right, stands the Saul of their ranks,—Uncompahgre peak, head and shoulders above all his comrades. Nor is this figure an idle comparison, for his tenon shaped apex easily suggests it.

Half way to our destination, the crazy buckboard rattles us painfully down a steep and stony hill into the valley of the Lake Fork of the Gunnison, where there is room for several ranches whose fields of hay and oats show a plentiful growth, and whose potato-patches are something admirable. The best of these is Barnum's, where there is also a store and a post-office, and where your "humble correspondent," supposing himself about to lay his head upon a soft bag of oats, nearly dashed his brains out by hurling it in misplaced confidence against a marble-solid bag of salt. *Eheu! miserere me!*

When we had wound our way farther up the narrow cañon into which the valley contracted on the further side of this gateway, there came to view the precise similitude (but here on a lesser scale) of the massive, pillared, mitre-crowned cliffs that form the shores of the Columbia river between Fort Vancouver and The Dalles.

As a mining-town Lake City is not now so active as formerly. It stands in a little park at the junction of the Lake Fork (of the Gunnison) with Henson creek,—both typical mountain streams, each wavelet flecked with foam and sparkling like the back of the trout it hides. Henson creek became especially famous among prospectors, who found that, however large an army of miners might flock in there, new veins were always to be had as the reward of diligent searching Thus a populous and highly enterprising town arose, which became the supply point for a wide mountain region, owing to its accessibility from both north and south; and though it was over one hundred miles—mountain miles at that!—from a railway, more than ten million pounds of merchandise, and five million pounds of mining machinery and supplies were taken in on wagons during 1880, at a cost of over a million dollars for transportation alone. A very good class of people went to Lake City, too, so that a substantial and pretty town arose, school-houses and churches were built, and I have never seen a mining camp where the bookstores and news-stands were so well furnished and patronized. At the beginning of 1881 about two thousand people lived in the town itself, not counting the great number of men in the mountains round about; and three factories for the treatment of ores were in operation.

Since then, however, Lake City has retreated somewhat; not that the mines have proved false to the confidence placed in them, but because it has been shown that until cheaper methods of transportation and more economic treatment can be devised, the mines cannot be worked to the same profit which a similar investment in some neighboring districts will return. This is due to the fact that the ores, of marvelous value when their mass is considered, are of too low grade, as a rule, to afford a high margin over the expenses of working. This by no means condemns the district; it only causes its stores of wealth to be held in abeyance for a while before their coinage. Many another district, a few years ago thought equally profitless, has risen to become the scene of steady dividend-making labor through the perfection of processes. It will not be long, before, by like means, the reviving of Lake City's mines will occur, and enable her to catch up with her more fortunate sisters in the wide circle of the San Juan silver-region.

But when that time has come, — though the Alpine grandeur of the scenery cannot be lost, the splendid shooting and fishing which now make the village one of the favored resorts of the west, will have disappeared; and there are some of us, more sentimental than world-wise, who will regret the change. Over these rolling uplands, among the aspen groves, upon the foot-

WINNIE'S GROTTO.

hills and along the willow-bordered creek deers now throng, and even an occasional elk and antelope are to be seen. In the rocky fastnesses the bear and panther find refuge, and every little park is enlivened by the flitting forms of timid hares and the whirring escape of the grouse disturbed by our passing. Upon these lofty, grass-grown plateaus, some cattle already get excellent feeding; and the time will be short before they are multiplied into the vast herds whose pasturage will be economised by good management, and for which a market will be found within a few days drive of the range. Too high and arid for extensive farming, the opposite, yet inter-dependent, pursuits of mining and cattle-raising, will ere long bring all this elevated interior of the state into full utilization. When one wonders how this railway company is to support itself amid the wilds, this future must be remembered.

XXVII

IMPRESSIONS OF THE BLACK CAÑON.

By what furnaces of fire the adamant was melted, and by what wheels of earthquake it was torn, and by what teeth of glacier and weight of sea waves it was engraven and finished into perfect form, we may hereafter endeavor to conjecture.
—John Ruskin.

T was with eager interest that we despatched a hasty breakfast, and attached our cars to the early morning express westward bound from Gunnison. The Grand Cañon of the Gunnison lay just ahead. An open "observation" car, crowded with sightseers, was hooked on behind us, but that did not interfere with our favorite rear platform, and thither our camp-stools were taken.

This river Gunnison has a hard time of it. The streams that finally unite to make it up, are loath to do so, and it came near not being born at all. The flat country we see just below the town vouchsafes a few quiet miles under the cottonwoods, but presently the hills close in, and then the river must needs gird up its loins for a struggle such as few other streams in the wide world know. Its life thenceforth is that of a warrior; and it never lays aside its knightly armor till the very end in the absorbing flood of the Grand.

Above the rattle of the train, echoing from the rocky highlands that hem it in, we can hear the roaring of this water as we thunder down its sinuous course toward Sapinero. Great fragments that have fallen from the steep banks, where an avalanche of stones lies precariously as though even the shock of our passing would set them sliding, fret the stream with continual interruptions and turn its green flood into lines of yeasty white. These same rocks are admirable fishing-stands, however, for the trout love the deeply aërated water that swirls about them; and we see more than one silvery fin snatched from its crystal home to hang in mute misery upon the angler's switch of forked willow.

"Do you think it's right?" asks the Madame, with a pitiful tone in her voice.

"No, but it can't be helped; and you'll find some casuistry to meet the case about dinner-time."

"Casuistry—casuistry?" says Chum reflectively. "Is that a new kind of sauce?"

Ahead the green hills, marked with horizontal lines, that we suspect to indicate outcroppings of lava, shut quite across our path. Nevertheless we can detect a dark depression toward which the track points straight as an arrow, and we suppose that at that point an entrance exists. Behind it stood summits so lofty that this barrier did not seem imposing; but now that a gateway has opened (yet far enough only for our track to enter by encroaching on the river's highway), we are surprised at the altitude of the walls which momently rise higher and higher on each side, as though we were descending a steep incline into

ECHO ROCK.

the earth. At what an abyss must the river lie in the middle of the range!

The early morning sun streams warm and rich into the cañon, dispelling the nocturnal chill and making the air delightful beyond expression. We are hurled along between close-shutting crags that are the type of solidity, yet seem to waver and topple at their summits as we gaze at them, cut strongly against the tremulous blue of the sky. Our ears are assaulted by the crashing of iron against iron and steam shrieking at the wind, and by the roar and dashing of enraged and baffled water. The lyric sweetness of the distant hill-picture caught in our backward glance as we entered the gates of the cañon, is gone; the poetry of this scene has the epic dignity and the stirring excitement of a war-story sung on the eve of righteous battle. This is the site and the monument of a struggle between forces such as we have no capacity to comprehend. Take a fragment of this shining rock not so large but that you may lift it, and you will find that studied ingenuity, and the vigorous application of power that men speak of as enormous, are required to break it into smaller pieces. Yet here are masses many hundreds of feet high and wide, that have been riven as I might halve a piece of clay. You may say it was done thus, or so. No matter, the impression of stupendous power remains and imprints itself deeply on the mind. Here for miles we pass between escarpments of rock, a thousand, fifteen hundred—ay, here and there more than two thousand feet high. This is not a valley between mountains with sloping sides slowly worn away. Here are vertical exposures that fit together like mortise and tenon; facing cliffs that might be shut against one another so tightly that almost no crevice would remain. To view this mighty chasm thoughtfully, is to receive a revelation of the immeasurable power pent up in the elements whose equilibrium alone forms and preserves our globe; and if we call it "awful," the word conveys not so much a dread of any harm that might happen to us there, as the vague and timorous appreciation of the dormant strength under our feet. If the gods we call dynamic can rive a pathway for a river through twenty miles of solid granite, of what use is any human safeguard against their anger?

But away with these serious thoughts! The cliffs are founded in unknown depths it is true, but their heads soar into the sunlight, and break into forms not too great for us to grasp. Straight from the liquid emerald frosted with foam which flecks their base—straight as a plummet's line, and polished like the jasper gates of the Eternal city, rise these walls of echoing granite to their dizzy battlements. Here and there a promontory stands as a buttress; here and there a protruding crag overhangs like a watch-tower on a castle-wall; anon you may fancy a monstrous profile graven in the angle of some cliff,—a gigantic Hermes rudely fashioned. In one part of the cañon where the cliffs are highest, measuring three thousand feet from the railway track to the crown of their haughty heads, faces of the red granite, hundreds of feet square, have been left by a split occurring along a natural cleavage-line; and these are now flat as a mirror and almost as smooth. On the other hand, you may see places where the rocks rise, not solid, sheer and smooth, but so crumpled and contorted that the partition-lines, instead of running at right angles, are curved, twisted and snarled in the most intricate manner, showing that violent and conflicting agitations of the rock must have occurred there at a time when the whole mass was heated to plasticity. In another place, the cliff on the southern side breaks down and slopes back in a series of interrupted and irregular terraces, every ledge and cranny having a shapely tree; while not far away another part of the long escarpment, the rocky layers, turned almost on edge, have been somewhat bent and broken, so that they lie in imbricated tiers upon the convex slopes, as if placed there shingle-fashion.

Just opposite, a stream whose source is invisible has etched itself a notched pathway from the heights above. It plunges down in headlong haste until there comes a time when there is no longer rock for it to flow upon, and it flings itself out into the quiet air, to be blown aside and made rainbows of, to paint upon the circling red cliffs a wondrous picture in flashing white, and then to fall with soft sibilancy into the river. The river has no chance to do so brave a thing as this leap of Chippeta falls from the lofty notch; but seeing a roughened and broken place ahead where the fallen bowlders have raised a barrier, it goes at it with a rush and hurls its plumes of foam high overhead, as, with swirl and tumult, and a swift shooting forth of eddies held far under its snowy breast, it bursts through and over the obstacle and sweeps on, conqueror to the last.

In the very center of the cañon, where its bulwarks are most lofty and precipitous, unbroken cliffs rising two thousand feet without a break, and shadowed by overhanging cornices,—just here stands the most striking buttress and pinnacle of them all,—Currecanti Needle. It is a conical tower standing out somewhat beyond the line of the wall, from which it is separated (so that from some points of view it looks wholly isolate,) on one side by a deep gash, and on the other by one of those narrow side-cañons which in the western part of the gorge occur every mile or two. These ravines are filled with trees and make a green setting for this massive monolith of pink stone whose diminishing apex ends in a leaning spire that seems to trace its march upon the sweeping clouds.

It was in the recesses of the rift beside Currecanti Needle, says a tradition which at least is poetic, that the red men used to light the midnight council-fires around which they discussed their plans of battle. Though judgment may refuse the fact, fancy likes to revel in such a scene as that council-fire would have made, deep in the arms of the rocky defile. How the fitful flashes of the pungent cedar-flame would have driven back the lurking darkness that pressed upon it from all sides! How, now and then starting up, the blaze-light would sally forth and suddenly disclose some captive of the gloom rescued from oblivion —perhaps a mossy bowlder, an aged juniper, a ghostly cottonwood stump, or a ledge of sleeping blossoms! How the bright and polished rocks would be re-reddened and sparkle at their angles under the glancing light; while the pretty soprano of the stream and the deep bass of the river's roar sang a duet to the narrow line of stars that could peep down between the cañon walls! Surely the time and place were suitable for planning the lurid warfare of a savage race; and as these untamed men, their muscular limbs and revengeful faces, disclosed uncertainly, like the creatures of a flitting fantasy, in the red firelight, enacted with terrifying gestures the fierce future of their plotting, a spectator might well think himself with fiends.

"On Night's plutonian shore,"

or else discard the whole picture as only the fantastic scenery of some disordered dream.

Opposite **Currecanti Needle and** cañon stand some very remarkable rocks, underneath the greatest of which the train passes. Then there is a long bridge to cross where the river bends a little; and perhaps the echoing chasm will be filled with the hoarsely repeated scream of a warning whistle. And so, past wonder after wonder, Pelion upon Ossa, buried in a huge rocky prison, yet always in the full sunlight, you suddenly swing round a sharp corner, leaving the Gunnison to go on through ten miles more of cañon, and crashing noisily through the zigzag cañon of the Cimmaron, which is so very narrow and dark it deserves no better name than crevice, quickly emerge into daylight and a busy station.

Thus I have tried to give the reader some trifling indication of what he may expect to see during his hour in the heart of the "Black" cañon, which is not black at all but the sunniest of places. I cannot understand how the name ever came to be applied to it. No Kobolds delving in darkness would make it their home; but rather troops of Oreades, darting down the swift green shutes of water between the spume-wet bowlders, dancing in the creamy eddies, struggling hand over hand up the lace ladders of **Chippeta Falls,** to tumble headlong down again, making the prismatic foam resound with the soft tinkling of their merry laughter. All the Sprites of the cañon are beings of brightness and joy. The place is full of gayety.

This sense of **color and light** is perhaps the strongest impression that remains. Though it is quite as deep and **precipitous as the** Royal gorge it is not so gloomy and frowning; though the cataracts are greater than those at Toltec, they are not so fear-inspiring. In place of dark and impenetrable walls, here are varied façades of lofty and majestic design, yet each unlike its neighbor and all of the most brilliant hue. The cliffs are architectural, suggestive of human kinship and more than marvelous—they are interesting.

Then there is the brilliant and resistless river. At **Toltec** it is only a murmuring cataract; in the Royal gorge a stream you may often leap across; the Rio de Las Animas is deep and quiet. But here rushes along its gigantic flume a great **volume of** hurried water, **rolled over and** over in headlong haste, hurled against solid abutments to recoil in showers of spray or to sheer off in sliding masses of liquid emerald. Now some quiet nook gives momentary rest. The water is still and deep. Small rafts of seedy foam swing slowly around the edges, tardy to dissolve.

GUNNISON'S BUTTE.

The rippled sand can be seen in wavy lines far underneath like the markings on a duck's breast. The surplus water curves like bent glass over the dam that rims the pool on its lower side, and beyond is a whirlpool of foam and the hissing tumult of shattered waves amid which rise the sharp crests of crimson bowlders flounced with snowy circles of foam.

Alternating with the vast pillars and the slick faces of red rock are the nooks and ravines where trees grow, flowers bloom and the eye can get a glimpse of a triangle of violet sky; while sometimes a silken skein of white water can be traced down the deepest recesses of the glen, and the gleam of swallow's wings flitting in colonies about their circular adobes bracketed against the wall.

These facts, noted like short hand memoranda upon the brain as they quickly flash by, slowly return to the memory and feed recollection, as the mind in after days elaborates the impression made by each, and summons a series of separated and leisurely pictures before the imagination; but no writer can depict for another what the form of these pictures shall be. I recite to you the elements—stupendous measurements, majestic forms, splendid colors, the gleaming green and white of water, the blue and gold of sun and sky, the crystalline sparkle of reddened rocks; but you must yourself receive these elements if you are to paint adequately to your fancy the pictures of the cañon. It is not literary cant, but the literal truth, when I say that to be understood, this marvelous pathway through the mountains must be seen. And having seen it, you have enriched your memory beyond anything you could have foretold.

XXVIII

THE UNCOMPAHGRE VALLEY.

*The hills grow dark,
On purple peaks a deeper shade descending;
In twilight copse the glow-worm lights her spark,
The deer, half-seen, are to the covert wending.*
—WALTER SCOTT.

HE station at the western end of the cañon of the Gunnison is called Cimmaron after the river upon whose banks it stands. In the prehistoric days before the railway, this was Cline's ranch, where all the stages from the Gunnison to the San Miguel region stopped. He was one of the few pioneers who got on well with the Indians, and his monument stands in the name of a peak down by Ouray.

From Cimmaron upward stretches one of the steepest grades between Denver and Salt Lake, in order to surmount Squaw Hill in the Cedar range,—the water-shed between the Cimmaron and the lower drainage of the Gunnison. Two locomotives drag us at a snail's pace, struggling, puffing rapidly and spasmodically just as though their lungs were tortured by the rarity of the air. Their efforts suggest Pope's line, and seem to beat time to it,—

"When Ajax strives, some rock's vast weight to throw,
The line, too, labors, and the words move slow."

There is nothing to be seen but great knolls of grass and sage brush, sometimes showing their rocky anatomy; and this nakedness is a relief after the strain upon our attention in the cañon. Finally we get high enough to look far away to a horizon full of hazy mesas and peaked mountains, with a touch of valley land down in the center of the picture. A cool breeze blows, and comes with refreshing.

The valley we see is our first view of the Uncompahgre; and in the middle of the afternoon we reach the town of Montrose,—a settlement of wooden houses.

Here we stopped. There were two reason: first, this was the point of departure for the upper valley of the Uncompahgre, and the mining region on the northern front of the San Juan mountains; second, we wanted to know the arguments that had induced some hundreds of people to make their homes in the midst of this white sahara.

The first of these objects required instant attention, for between our arrival at Montrose and the departure of the stage up the valley to the Uncompahgre Cantonment, and the town of Ouray, there was time only to get a hearty luncheon. Chum had said from the start that he was quite willing to concede all the attractions of Ouray, and declined positively to leave the comfort of home. I told him he was missing a good deal, but he said that he had lost all faith in good deals—didn't "gamble any more on that chance,"—and persisted in his "No, thank you."

The Madame felt both inclined and disinclined. She knew the horrors of staging, she said it was a fit punishment for malefactors, and she

BUTTES OF THE CROSS.

dreaded even forty miles of it, on a level road, worse than a fit of sickness. Then she looked unutterable sympathy at me, and began to reflect that possibly her duty as a wife required her to go (seeing that I couldn't escape it,) in order to share the discomforts her husband was obliged to undergo, and do what she could to alleviate his tortures.

Just at this juncture, for doubt had swayed her usually well-decided mind up to the last minute, she caught a glimpse of the big red coach coming from the hotel toward us. Its noise was as the thundering of "the chariots of Israel and the horsemen thereof." It swung from side to side like a fire-steamer tearing over Baltimore cobble stones. It lunged into irrigating ditches and came pitching up out of them, while the hind boot dived in to be brought up with a frame-cracking jolt, and it rocked fore and aft like a Dutch lugger in a chop sea. Its great concavity was packed full of unfortunate Jonahs, swallowed "bag and breeches." Its capacious baggage receptacles before and behind were distended with trunks and valises, rolls of blankets, packages of newspapers, boxes of fruit and dozens of mail-sacks. Its roof was piled with a confused mass of luggage and sweltering humanity. There wasn't a lady to be seen. Chum looked at me as I buttoned my duster, and lifting a corner of the Madame's apron to his eye, choked back a sympathetic sob.

"Come on!" I called to the person who was anxious to alleviate my tortures, but she held back.

"I'm—thinking—whether—after all"—

"Oh, are you? Good—give us a kiss—good bye! Better do your thinking now than after you're tired out up the valley. I'll be back shortly, and expect you to know all about Montrose."

The big red coach came to a lurching anchorage close by the door and I climbed to the vacant seat beside the driver, for which, with the wisdom of experience, I had telegraphed a request the day before.

"Been a-keepin' this seat for you with a club," said Jehu, curtly, as he gathered up the reins of four grey horses and removed his foot from the brake.

There was the sensation of a geological upheaval, forward. I dug my heels hard into the mail-sacks heaped upon the foot-board, clutched the hand-rail of the seat, set my back against the knees of the man on the dicky seat, stiffened my neck to save my head from being snapped off,—and we were under way.

A whole chapter could be written about that stage ride and my fellow travelers, but it will keep. The road crossed the yellow Uncompahgre, and stretched like a chalk-mark athwart the sage green expanse of gravelly valley. One of the outside passengers was an Englishman who had spent seven years in the diamond fields of South Africa. He told us this region reminded him of that land, and entertained us by his accounts of it, and of the Caffres—especially the English habit of knocking one down (a Caffre-boy was always handy) whenever the aggrieved temper of a white man required any little relief. So, with umbrella overhead and green goggles to break the glare,—despite the purple-blue storms we could see stalking about the mountain-ranges ahead—the first seven miles passed speedily, and we drew up at the sutler's store of the pretty military cantonment, whose buildings had loomed mirage-like for more than half the way.

When the Utes were ready to be moved from this valley over to the new Uintah agency, a military post was established here. As its permanence was not decided upon, only a cantonment was founded, providing temporary quarters in log houses for six companies. It was called simply the Cantonment on the Uncompahgre, and was at first garrisoned by the Twenty-third Infantry. In 1882, however, this regiment was relieved by four companies of the Fourteenth Infantry. Its commanding officer, Lieutenant Colonel Henry Douglass is here, although the headquarters of the regiment are at Sidney, Nebraska.

The post, as I said, is not intended to be a permanent one. The visitor must not expect, therefore, the handsome buildings and grounds to be seen at Camp Douglass, Salt Lake City, where this regiment was domiciled for seven years, enduring meanwhile some hard service in Indian fighting.

We had already passed Ouray Farms, where Ouray, the fine old head-chief of all the Ute confederation, lived toward the end of his life in a good house built of adobe after the Mexican fashion, and cultivated the neighboring bottom-lands. His farm made the grand center of Ute interest, and from the pleasant groves near it radiated all the trails across mountain and plain. Many out-houses, of log and frame, surrounded the main building and testify that order was one of the great chief's good qualities. Here, after his death Chippeta, widow of Ouray, continued to live, raising farm-products and pasturing sheep, and so attached had she become to the spot that she importuned the government to be granted the privilege of abandoning her race and returning to her farm home. The government refused this request, but decided to sell the farm for the personal benefit of Chippeta.

All along the river, which ran between thick belts of trees some distance at our left, we had seen spaces of meadow and a few ranches. At the old Agency—four miles above the post—we came to its lofty bank at a point where the river bent in so close to the foot of the bluff, that water for the station-stables was drawn up by means of a pulley mounted in a tall scaffolding of poles standing in front of the cliff, and reached by a bridge. It was a well, built some sixty feet out of ground, like that Nevada tunnel Mark Twain describes, which, having gone quite through the hill, was continued out upon a staging.

The road thence ran along the edge of the bluff and through the woods—a road upon which we rattled at a steady trot, although on the left there was nothing to break our fall for a hundred feet or so, should an accident tip us over. The river valley, thus sunken, was sometimes narrow and the stream turbulent among rocks; sometimes a mile or two wide with willow-covered bottoms; sometimes showing islands crowded with trees and thickets, or of great bends where lay spaces of rank meadow. Two or three little houses were pointed out where head men among the Indians had lived on small farms, and the driver, who had run a stage before the red men left, told us many interesting stories of their life in this favorite valley.

Leaving the river and the verdant gorge, its cottonwoods illumined with flaming light of the sunset, the road took to the higher ground and gave us many a good jolt in crossing the small accquias which watered the upper ranches on the edge of the mesa, and then came into view of Uncompahgre park, stretching away to the westward like a prairie, and the scene of some of the finest farming in Colorado. There are about thirty ranches where, half a dozen years ago was nothing but wild pasture. The ranchmen were all poor men when they came here; now they have pleasant houses, well fenced and irrigated farms and equipments in abundance. I heard of one ranch sold lately for ten thousand dollars, and was told of another where the owner cleared six thousand dollars for his last season's profits. Everything is raised except Indian corn, but wheat is not cultivated so extensively as it will be when milling facilities are better. Barley, oats, hay and vegetables are the principal crops, and potatoes probably offer the highest return of all. Prices have decreased to one-fifth the figures of five years ago, yet the ranchmen prosper and increase their acreage, putting surplus money into cattle which roam upon the adjacent uplands. The land is by no means all taken up, however, and improved property can be bought at reasonable prices. There is plenty of water, too, an important consideration.

In the center of the park we passed a copious spring of hot mineral water, carrying much iron, as we could tell by the circular tank of ferric oxide it had built around it, forming a bath large enough for a hundred persons at once. As yet there are few arrangements for making use of this fountain,—a fact due to the plentiful hot springs of iron and of sulphur (sulphate of lime, etc.), water close to Ouray, where a sanitarium and bath houses have been fitted up, and where persons suffering from rheumatism and kindred ailments find great benefit. So much warm water is poured into the Uncompahgre, in fact, that nothing more than a film of ice forms upon it in the coldest weather. Remembering all these varied advantages, it is no wonder the Utes loved the place and protested against its loss.

The mountains ahead came into plainer view, as we left the park; we caught a glimpse of the curious Sawtooth range off at the left, saw that the rounded outlines of the bluffs on each side were changing to abrupt walls, and trending inward, and then the hush of night and the quiet of weariness came to still our conversation and turn our thoughts into meditative channels. Darkness enveloped the world and we pulled slowly through it by the light of a thousand brilliant stars— the same stars that shone on the Madame and Chum; that, beyond the Range, shed soft light on the shepherds and herdsmen of the great plains; that trembled in the eddies of the Mississippi; that were watched by wakeful people on the slopes of laurel-crowned Alleghanian hills; that caught faintly the eye of revelers—for it must now be after the opera—in New York; that spoke a mysterious language to the watcher upon the far ocean; and, oh, best of all! that looked in at a curtained New England window and saw a child in peaceful slumbers. Little daughter under the ancient elms,—planet in the far sky,—father passing under the massive shadow of gigantic cliffs whose pine-fringed bulwarks are lost in the thick gloom above! What an immeasurable triangle, yet how swiftly does the mercury of thought compass it and link its points together?

XXIX

AT OURAY AND RED MOUNTAIN.

Bathed in the tenderest purple of distance,
Tinted and shadowed by pencils of air,
Thy battlements hang o'er the slopes and the forests,
Seats of the gods in the limitless ether,
Looming sublimely aloft and afar.
Above them, like folds of imperial ermine,
Sparkle the snowfields that furrow thy forehead,—
Desolate realms, inaccessible, silent,
Chasms and caverns where Day is a stranger,
Garners where storeth his treasures the Thunder,
The Lightning his falchion, his arrows the Hail.

—BAYARD TAYLOR.

URAY is—what shall I say? The prettiest mountain town in Colorado? That wouldn't do. A dozen other places would deny it, and the cynics who never saw anything different from a rough camp of cabins in some quartz gulch, would sneer that this was faint praise. Yet that it is among the most attractive in situation, in climate, in appearance, and in the society it affords, there can be no doubt. There are few western villages that can boast so much civilization.

Ouray stands in a bowl-shaped valley—a sort of broad pit in fact—hollowed out of the northern flank of that mass of mountains which holds the fountains of so many widely destined rivers. A narrow notch in the bowl southward lets the Uncompahgre break through to the lowlands, and furnishes us with a means of ingress; otherwise the most toilsome climbing would be the only way to get into or out of town. From this point diverge three or four short but exceedingly lofty, and several lesser ranges, like the spokes of a wheel from its hub. Eastward stretches the continental divide of the Sierra San Juan proper; southward the Needles and the circling heights that enclose Baker's park; westward the Sierra San Miguel; northward the spurs of Uncompahgre; and the diminishing foothills and mesas that sink gradually into the Gunnison valley.

Yet the first comers—it is only seven years ago, but the mists of antiquity seem to gather about it—did not enter that way, but came over the range from the south. Prospectors for precious metals, they ascended the Rio Las Animas from Baker's park, until they found its head, and stood upon the dividing crest of the range. Here a streamlet trickled northward, and they followed its broadening current down the unknown gorges into which it sank. The walls were often too steep to allow any foothold for them, and then they would wade in the icy water and stumble over the slippery bowlders that had fallen from above. When a dozen miles of this work had been accomplished, they found themselves entering a cañon so narrow, that by stretching out their arms they could almost touch both of its walls; and so irregular that a few rods before and behind was all the distance that ever could be seen at once. Uncertain when they would be brought to a standstill by some pool or precipitous fall, and compelled to struggle back against a torrent which scarcely allowed them to move downstream in safety, they pushed on until they suddenly emerged into a beautiful round valley, filled with copses of trees and sunny glades. In this haven the chilled and weary prospectors rested for the night. While one man—there were no more than three, I believe,—built the fire, sliced the fat bacon and molded the bread; the second went to the river with his fishing-line,

MARBLE CAÑON.

and the third started out with his gun. By the time the bread was baked the angler came back with eighteen trout and the hunter returned for help to bring in a bear he had killed within a couple of hundred yards. So runs the tradition, and there is no reason to discredit it.

Now, where the bear was shot and the trout caught, stands a town of fifteen hundred people, which forms the center for supplying a wide circuit of mining localities, including Red Mountain, Mt. Sneffles, Mineral Point and Mineral Farm, Bear Creek, and half a dozen other places of lesser note; and which affords a good market for the agriculturists of the lower valleys, and the cattle breeders of neighboring mesas. Prosperity, comfort, and even much luxury prevail now; but some of the trials of the earliest settlers, beset by isolation, winter, famine, and the fear of Indians, would be worth recounting could I have unlimited space.

This is not a miner's guide, and nothing could be drier reading for a stranger than a catalogue of diggings and minerals. The ores abound in a thousand ledges which run up and down, and here and there, all through the mountains, from the metamorphic limestones of the outer ledges to the storm-hewn trachyte that caps the hoary summits. What I have said concerning the ore of the opposite (southern) side of the San Juan system of mountains, and the way in which it occurs, applies well enough to this side also. It could not well be otherwise, for the age and general geology of the two regions is as nearly alike as the two sides of the same mountain-chain are very likely to be. In a word the ores are varied, but chiefly ores of galena and copper, occurring in fissure veins and carrying a "high grade" proportion of silver (in various forms) and a considerable quantity of gold. The extraordinary variety of minerals, and the vast bulk of the ore deposits are the two noteworthy features of the region. These ores, moreover, as a rule, are not "refractory" though containing antimonial elements which in an excess would make them so. Works for their concentration, *i. e.*, the sifting out (after pulverization) of the worthless vein-matter, in order to save the expenses of transportation, are run to great advantage.

Ouray's principal claim to our notice as sightseers lay in its beautiful situation, and the attractive bits of mountain scenery in its neighborhood,—a collection of pictures which it would be hard to duplicate in an equally limited space anywhere else in the whole Rocky mountains.

The valley in which the town is built is at an elevation of about 7,500 feet above the sea, and is pear-shaped, its greatest width being not more than half a mile while its length is about twice that down to the mouth of the cañon. Southward—that is toward the heart of the main range,—stand the two great peaks Hardin and Hayden. Between is the deep gorge down which the Uncompahgre finds its way; but this is hidden from view by a ridge which walls in the town and cuts off all the farther view from it in the direction, save where the triangular top of

Mt. Abrams peers over. Westward are grouped a series of broken ledges, surmounted by greater and more rugged heights. Down between these and the western foot of Mt. Hayden struggles Cañon creek to join the Uncompahgre; while Oak creek leaps down a line of cataracts from a notch in the terraced heights through which the quadrangular head of White House mountain becomes grandly discernible,— the eastermost buttress of the wintry Sierra San Miguel.

All of these mountains, though extremely rugged, precipitous, and adorned with spurs and protruding shoulders of naked rock, yet slope backward somewhat, and through one of these depressions passes a most remarkable and picturesque wagon road to Silverton, constructed at immense cost and displaying wonderful engineering skill. But at the lower side of the little basin, where the path of the river is beset with close cañon-walls, the cliffs rise vertical from the level of the village, and bear their forest-growth many hundreds of feet above. These mighty walls, two thousand feet high in some places, are of metamorphic rock, and their even stratification simulates courses of well-ordered masonry. Stained by iron and probably also by manganese, they are a deep red-maroon; this color does not lie uniformly, however, but is stronger in some layers than in others, so that the whole face of the cliff is banded horizontally in pale rust color, or dull crimson, or deep and opaque maroon. The western cliff is bare, but on the more frequent ledges of the eastern wall scattered spruces grow, and add to its attractiveness. Yet, as though Nature meant to teach that a bit of motion,—a suggestion of glee was needed to relieve the sombreness of utter immobility and grandeur however shapely, she has led to the sunlight by a crevice in the upper part of the eastern wall that we cannot see, a brisk torrent draining the snowfields of some distant plateau. This little stream, thus beguiled by the fair channel that led it through the spruce woods above, has no time to think of its fate, but is flung out over the sheer precipice eighty feet into the valley below. We see the white ghost of its descending, and always to our ears is murmured the voice of the Naiads who are taking the breathless plunge. Yet by what means the stream reaches that point from above, cannot be seen, and the picture is that of a strong jet of water bursting from an orifice through the crimson wall and falling into rainbow-arched mist and a tangle of grateful foliage, that hides its further flowing.

As Mr. Weston well says, and as I have insisted in my chapters upon the southern side of the San Juan range, the indescribable charm of this scenery is due not so much to its gigantic proportions, its grotesque and massively-grand outlines, or its variety of composition, as to the contrasts of color and condition. "Even now (May) while I write," he says, "it is warm and summery in town, the side hills are covered with flowers and the whistle of the humming bird's wing is heard in the air; yet I can look up at White House peak and see the snow banners blowing from its summit, as in the coldest day in winter. In the autumn, more especially, are the contrasts of color seen, and the landscape, as it then appears, if painted on canvas, would, I believe, be laughed at, if shown in Europe or the Eastern States, as an impossibility. I have climbed the heights above Ouray, and looked down on it, when the atmosphere of the valley seemed of a hazy blue, the sloping sides of the surrounding mountains being clothed with the golden yellow and the red brown of the quaking aspen and the dwarf oak, the varied greens of the spruce, balsam, cedar and yellow pine, and above that the brown gray of the trachyte peaks, their snow-capped summits forming a charming contrast against the lovely violet blue of the evening sky."

This valley alone, with its everchanging panorama of summer and winter, of verdurous spring and the noise of gushing waters, of flaming autumn and the drapery of haze etherializing the world, presenting under always novel aspect the forms and colors so lavishly displayed— this nook alone would satisfy a generation of artists. But the enchantment of the half hidden gorges, the allurement of the beckoning peaks urge us to explore beauties beyond.

I cannot redescribe the way in which these bristling peaks of purple and green trachyte cut the tremulous sky, nor try to make you understand anew the abysses that sink narrowly between the closely crowded mountains. If the reader will kindly turn back to where I have endeavored to convey to him some idea of the Alps that lie about Baker's park and at the head of the Rio Dolores and the Rio de La Plata, he will learn what I might repeat of scenes this side of the divide; for some of those former peaks can be seen from here, and this, too, equally with the southern slope, "is Silver San Juan."

The ride across the hills towards Red mountain was something to be remembered The great walls of maroon rock and the precipices that rose in terraced grandeur upon their shoulders, coming into view one by one as we ascended from the basin to the foothills, were all wet with the night dews, and gleamed like mirrors under the morning sun. The foothills themselves were rugged jumbles of rocks heaped about the base of the mountains, and full of deep crevices where the streams coursed far out of sight and hearing. They were covered with a mingled growth of spruces, cedars, small oaks and several other shrubby trees. There were open spaces where a dense chapparal or heather of small thorny bushes of various kinds hid the ground; and other slopes where tall grass and innumerable flowers formed favorite pastures for sedate groups of donkeys. Passing the dizzy brink of the chasm into which Bear creek makes its awful leap, snatching a beauty beyond portrayal from the very jaws of terror, we enter a rank forest of aspens and spruces. One might fire a pistol-ball across to the side of Mt. Hayden, which rises an almost sheer wall of indigo-gray from a gulf between

GRAND CAÑON OF THE COLORADO.

us and it, whose creviced-bottom is out of sight below. Deep and varied shadows lie in the little ravines that seam its almost vertical slopes, and streaks of dusty snow lurk in the higher crannies feeding trickling cascades of sunbright water that drop like tears into the unfathomed cañon.

Through the trees southward, to the right of the triangular peak of Engineer mountain, and the great barrier of Abrams, we could now catch a glimpse of a rounded summit as gaudy as the hat of a cardinal. This was the Red mountain, of which so much has been heard. The road there follows the course of Red Mountain creek from its mouth for two miles through dense pine timber. At this point, four miles from Ouray, and two thousand feet higher, it enters a flat valley or park two miles long, which is covered with willows and with prairies of long grass that every autumn is mowed for hay. This park contains many ponds and miry places, and is said to be underlaid everywhere with bog iron-ore. On either side of the park is a high range of mountains and trachyte peaks, that on the west being the divide between the Red Mountain district and Imogene basin in the Sneffels district, and that on the east being the divide between the Red Mountain district and the Uncompahgre district and Poughkeepsie gulch. At the upper end of the park commences the chain of scarlet peaks, from twelve to thirteen thousand feet in altitude, which are regarded as the volcanic center toward which all the lodes of the surrounding region seem to converge.

The history of this new "camp," Red Mountain, is a short one. In the summer of 1881 three men discovered the Guston mine, but as the ore was low grade it was worked only because it gave an excess of lead which was just then in demand at the Pueblo smelter. In August, 1882, John Robinson, one of the owners, was hunting deer, and while resting, carelessly picked up a small bowlder, after the manner of prospectors who never stop a moment anywhere but they scrutinize every bit of stone within reach, out of pure habit. Astonished at the weight of this piece he broke it in two and found it to be solid galena. This clue led to the discovery of the Yankee Girl lead close by. A month later the owners had sold the prospect-hole for $125,000,—but retained two other apparently equally valuable mines near at hand. In the Yankee Girl rich ore was found only a dozen feet below the surface; and though it had to be packed upon mules and burros all the way down to Silverton, it yielded a profit of over fifty dollars a ton.

Upon the heels of this discovery there was a great rush of miners and speculators toward the scarlet heights, and several large settlements—principally Ironton and Red Mountain Town—sprang up on the rough and forested hillside. Claim stakes dotted the mountain as thick as the poles in a hop-field, and astonishing success attended nearly every digging. Among them all the first lode opened, the Yankee Girl, held supremacy, as is so often the case; but a few months later a neighboring property, the National Belle, leaped far to the front at a single bound.

This occurred by the accident of a workman breaking through the tunnel wall into a cavity. Hollow echoes came back from the blows of his pick, and stones thrown were heard to roll a long distance. Taking a candle, one of the men descended and found himself in an immense natural chamber, the flickering rays of the light showing him the vaulted roof far above, seamed with bright streaks of galena and interspersed with masses of soft carbonates, chlorides and pure white talc. On different sides of this remarkable chamber were small openings leading to other rooms or chambers, showing the same wonderful rich formation. Returning from this brief reconnoisance a party began a regular exploration. They crept through the narrow opening into an immense natural tunnel running above and across the route of their working drift for a hundred feet or more, in which they clambered over great bowlders of pure galena, and mounds of soft gray carbonates, while the walls and roof showed themselves a solid mass of chloride and carbonate ores of silver. Returning to the starting point they passed through another narrow tunnel of solid and glittering galena for a distance of forty feet, and found indications of other large passages and chambers beyond. "It would seem," cries the local editor in his account of this romantic disclosure, "as though Nature had gathered her choice treasures from her inexhaustible storehouse, and wrought these tunnels, natural stopping places and chambers, studded with glittering crystals and bright mineral to dazzle the eyes of man in after ages, and lure him on to other treasures hidden deeper in the bowels of the earth. . . The news of the discovery spread like wildfire, and crowds came to see the sight, and to many of them it was one never to be forgotten."

This was only the first of these surprises, for many cavities have since been divulged, great and small, in each of which crude wealth had been locked up since the world was made. The character of the ores, the occurrence of these cavities, and the extremely short distances beneath the turf at which rich ore is struck, have combined to cause much discussion among geologists as to the true history of the district.

One of the most striking scenes in the neighborhood of Ouray is the passage through which Cañon creek makes its way down to join the Uncompahgre just above the village. A wild and interesting gorge leads upward toward the western foot of Mt. Hayden, the trail carrying one along the edge of a little cliff, and the walls rapidly contracting so that little room is left even for the foot-trail. A quarter of a mile, perhaps less, above the village, these walls suddenly close together, and the steep, brush-grown slope, is lost in a lofty crag, towering far above the tallest spruces, and standing squarely across the gorge. In this escarpment a zigzag crevice shows itself extending from top to bottom: at the top you may look some distance within it, but at the foot the protruding masses on one side, the sharp curve on the opposite, and the deep shadows, never illumined by the highest sun, shut off all searching by the eye. Out of this narrow, upright, cave-like crevice, as though from its original strong fountains, gushes the deep and turbulent stream, cold as ice and sparkling with a million imprisoned bubbles of air. Get as near as you can to its aperture—crane your head around the very corner of these mountain water-gates, and you can see nothing but darkness, in which only the outlines of the nearest irregularities in the rocky walls are dimly defined, while the beetling ledges above shut out the narrow line of sky that might be seen were the sides of the cañon smooth. Retreating down stream a little way, you look past bright pillars of rosy quartzite, across the glittering pathway of foam flecked water, glorying in its escape, up to the lofty gates and the shadowy crevice between, whence the river comes ceaselessly and with singing; you note the color-touches of the flowers and blossoming vines; the soft hangings of the ferns under the damp ledges, the emerald foliage of the poplar standing bravely beneath the shadow of the cliffs and the darker forms of giant spruces—you see this contrast of brightness and color and sunshine just without the damp gloom of the mysterious portals; and you tell yourself that there are few scenes in the Rockies that can equal it.

There is a roundabout way to get to the top of these cliffs and look down into the chasm; and at one point, where it is much more than one hundred feet in depth, a person may easily step across from edge to edge. Though it would probably be impossible in the lowest stage of water to make one's way up from below against the swift flood that fills the whole width of the chasm, yet by going above it is possible to work your way down stream for a long distance into the crevice. A cave exists there, entered at the surface of the water, and occasional picnic parties are made up to go to it. These consist mainly of young people whom age has not sobered, for during the latter part of the way it is needful that the gentlemen wading should carry the ladies across frequent portages—to borrow a word from a reverse custom. The cave entrance at the water side is only an ante-chamber to the real cavern. To reach that a ladder and rope is required, by which the men first ascend to a second higher chamber and then draw the ladies up. The entrance is a hatchway so narrow that portly persons have been known to express fears as to their getting through.

Both cave and cañon are eaten out of the limestone, and several chasms of the same sort occur upon this and neighboring streams, where the water flowing along the strike of the upturned strata, has cut into it a narrow channel between walls of more resisting rock. Along Portland creek, just above the village, such a cañon is to be visited, containing many beautiful cascades, where the cañon walls do not rise vertically but at a considerable slant, one leaning over the other, and the stream ever edging sidewise as it cuts deeper and deeper. The erosion in these cases is not accomplished so much by attrition, as by a chemical decomposition of the limestone. Yet attrition must do a great work at times; for now and then these purling brooks become the channels for

GRAND CAÑON, FROM TO-RO-WASP.

cloudbursts at their sources, and then a mighty and impetuous flood hurls itself down the gorge and chokes the bursting cañons with an unmeasured mass of water and detritus, whose weight and velocity are so great, however, that the flood of water not only, but thousands of tons of bowlders and rocky fragments are forced through and spread out in the valley below. Every such a deluge leaves its marks plainly upon the sides of the cañons, as well as upon the softer banks outside.

It was in the afternoon that I mounted the coach homeward bound, and bade good bye to a host of pleasant acquaintances. I felt rather guilty. I had stayed longer than I intended, and had had a much better time than I had anticipated. I felt somehow, therefore, as though I defrauded the Madame and Chum out of a pleasurable opportunity; and I resolved to note more carefully whatever I might see that was delightful.

The gap in the great red cliffs which lets the river and us out to the lowlands is only two or three hundred yards in width, and is filled with a dense growth of trees. The river trickles as a narrow, winding stream through a broad swath of pebbles that it has swept down, and which annually it overflows. The lofty cliffs stand on each side, erect and imposing. Theirs were the massive forms seen dimly in the darkness and enveloping us in inky shadows when we came up at midnight. Their irregularities break into new forms of picturesqueness as we roll past, enchanting our eyes. Three or four miles below town the thick growth of trees in the bottom and on the ledges thins out, while the valley grows wider, and ranches begin to appear. Pleasant houses, surrounded by trees, stand in the midst of wide fields of grain and low-lying meadows of natural hay. The cliffs still rise hundreds of feet high and are redder even than those above,—as brightly vermillioned as the crest of the treasure-mountain I have compared to a cardinal's hat. Those on the eastern side (we are heading squarely northward) are sparsely wooded with spruces and cedars that get a foothold on the rocky shelves and lean outward craving the light; those on the left are almost bare, even of herbage.

It is said that *Uncompahgre* in the Ute tongue, means "red stream," and, if so, it is easy to understand the application. The water is not red (though it might sometimes look so when roiled by freshets,) but the whole cañon is crimson and blood-stained. The color shines between bushes and trees, stands out in great upright masses, tinges the dust underfoot, and intensifies both the green of the heathery hills and the azure of mountain and sky.

At Dallas Station, where the Dallas river comes down from the west into the Uncompahgre, we stopped to get supper and wait for the stage from Telluride, Rico, Ames, and the other mining towns in the San Miguel range, whose outlet is this way. Those mountains were in plain view—Sneffles, Potosi, and all their peers,—glorified in the sunset;

while away in the eastward could be seen the gashed and splintered peaks of that quartzite group (here called the Sawtooth, but reckoned on the maps as part of the Uncompahgre range) the outline of which I can only compare to the jagged confusion of the broken bottles set along the top of a stone wall.

It is dark when we leave Dallas and darker in the gloom of the mesa shadows and under the shrubbery that overhangs the road along the high river bank. Out of the blackness below came up the sound of the river's fretting as from a nether world, for we could only now and then get a glimpse of the shaded water. When this had been passed, however, and we were going at a steady trot across the wide-reaching and starlit uplands, it was very delightful. The air was cool and soft and drowsy. The stars shone with that brilliance which long ago suggested to the savage mind that they were pin-holes in the canopy through which beamed the ineffable refulgence of an endless day to be attained when the probation of this life was over. Every moment or two a meteor would leap out, flash with pale brilliance across the firmament, eclipsing the steady stars for an instant, and then disappear as though behind a veil. Sleepy cattle, resting in the dust, would rise with heavy lurchings to get out of our way and stupidly stare at us as we swung past. The "watch dog's honest bark" came to our ears from ranches, whose position we knew by a dot of yellow light; ghostly forms would quickly resolve themselves into the white hoods of freight wagons, their poles piled with harness and their crews asleep underneath; faint rustlings in the sage-bush told of disturbed birds and rabbits; and so, peacefully and enjoyably, midnight brought us to our journey's end.

XXX

MONTROSE AND DELTA.

*My father left a park to me,
   But it is wild and barren,
A garden too with scarce a tree,
   And waster than a warren;
Yet say the neighbors when they call,
   It is not bad but good land,
And in it is the germ of all
   That grows within the woodland.*

—TENNYSON.

HE compassion I had been feeling for probable *ennui* endured by the two who had been left behind at Montrose was quite unnecessary. They had amused themselves very well during my prolonged absence.

"Montrose is better than it looks," they told me.

"But what did you do?" I asked.

"Well, we studied the situation," said Chum, who is becoming thirsty for knowledge in these latter days. "And we got acquainted with some very pleasant people, who told us good stories, and took us out riding and lent us books."

"Yes," said the Madame, "in one of our rides we went up to the camp."

"Eh!"

"And heard how you spent a whole day there doing nothing but playing billiards with the officers. Do you call that being industrious?"

"Well," I began.

"No, it is not well at all; at any rate you might have told me, and not made believe you only saw the camp by passing through. And we heard all about that hop in Ouray. You forgot that, too, didn't you? The people were greatly surprised to learn you were not a gay young bachelor. It was three o'clock in the morning before you went home."

"Oh, Oh-h!" groans Chum.

"'Pon honor, it wasn't," I protest. "It was only two."

"Only two! Well, the next time you go to Ouray I'm going with you."

Chum sings:

"Now is the time for disappearing,' and takes a header out of the side door. It is my cue to follow suit, but instead the Madame picks up her parasol and sails out with dignity. She wouldn't make a bad figure in the Lancers, I think, as she closes the door. I had intended to do some writing before the time came to pursue our journey, but I don't feel like it now and pick up *Felix Holt* and a cigar. Presently the two return in high good humor over some joke, and luncheon is ordered and eaten amid a fusilade of chattered nonsense.

Betweentimes I extract bits of information in regard to Montrose and its neighborhood. The town is the center of a very large agricultural district. It supplies all of the valley of the Uncompahgre as far south as Dallas creek, and westward nearly to Delta; while northward its bailiwick extends over to widely scattered but numerous settlers on the North Fork of the Gunnison and its tributaries. A glance at the map will show the reader that a great number of small streams come down from the mountains lying north of the Gunnison, and of its North Fork, to feed those trunk-streams. The mountains themselves, and the spurs that stretch down between the creeks are rocky, sterile, and too cold for farming; but in the valleys where the descent is always rapid, water is easily led aside in irrigating ditches, and the soil is invariably found to be rich and responsive. Throughout these remote creek-sides, then, a large farming and stock-raising population has already settled itself; and though out of sight, it forms a large element in the class of buyers for whom the merchant at the railway station must provide. Those living on the lower part of the North Fork trade at Delta.

Lately coal has been found within half a dozen miles of town, and veins of great thickness and soundness are being opened in several places by Montrose men, who can sell it much cheaper than it has hitherto been brought from Crested Butte. At Cimmaron, about twelve miles from Montrose, coal of very good quality occurs in great abundance, and is being mined. On the mesas, surrounding Montrose, grows timber of unusual size and importance, and nearly all the large sticks—some forty-four feet in length,—used by the railway in the construction of bridges on this half of the line, were derived from those forests of yellow pine. Several sawmills, each a nucleus of small settlements, buy and sell at Montrose. Local cattle-owners make the town their headquarters, the herds ranging on the upland pastures within a few miles. The cattle business in this region has just begun, but everything proves so favorable that great expectations are entertained of it as a source of wealth. The object is to raise fat beef for local markets, and Durham blood is being introduced to raise the grade of the native stock. The Cimmaron range, the heights beyond the North Fork and the Uncompahgre mesa, supply the chief ranges at present. A good many people are employed at Montrose, also, in the forwarding business,— that is, the re-loading of merchandise and other goods into the huge trailed wagons which they used to call "prairie schooners" on the plains, to be dragged away to the mountain mining camps. Finally, the town is the county seat.

EXPLORING THE WALLS.

While these resources are all of importance Montrose depends mainly upon the farming which she says is to make her valley and the dun-colored mesas, "blossom as the rose."

"They tell me," says Chum, "and they prove it, too, that there is nothing you cannot raise here short of tropical fruits, and they're not quite sure about that, for they propose peaches, nectarines, and apricots. And as for grain, great Injuns! why I saw stalks of oats as big as a walking-stick, and stems of barley that looked like gun-barrels."

The Madame raises her eyebrows and coughs slightly, but I take no notice.

"And as for wheat, sir,—wheat? why it's immense! Thirty-five and forty bushels to the acre is the regular yield, and of oats they will produce fifty or sixty bushels, and of barley eighty or ninety. As for corn, I forget the figures, but when we go down the road this afternoon you'll see great green fields of it that'll make you think you're back on the banks of the Wabash. There isn't anything they can't raise in these bottoms, where they have more water than they know what to do with, and it'll be only a few years before this whole great patch of greasewood and chalk will be verdant with—with potatoes and corn."

It was a bit of a break, but when this young man gets a fair grip upon poetry he don't let go so easy. He frowned down the suspicion of a smile round the corner of our eyes, and rising to his feet, continued:

"I tell you, sir, in five years from now the people of this favored spot can say in the words of the immortal singer—speaking historically, of course, you understand—can say,

> "Behind, they saw the snow-cloud tossed
> By many an icy horn"

*   *   *   *   *   *   *   *

> "Before, warm valleys, wood-embossed
> And green with vines—

(watermelons, squashes, pumpkins, hops, morning glories, grapes, strawberries, parsley, honeysuckles—I've seen 'em all!)

> "—and corn."

We exploded with laughter, and even the enthusiastic orator smiled grimly as he sat down.

"May be Mr. Whittier wouldn't have seen so much poetry in the way I used his words, but I tell you Montrose knows there's a heap of truth in it."

"Yes, no doubt. But how about the 'icy horn'—these high and dry benches up here?"

"Well, they say the very strongest and most productive soil of all is on those same gravelly mesas. It's lighter and different from the saline clays of the bottoms. Now, over there"—pointing to the great upland, which lay elevated a hundred feet or so above the river on the southern side of the Uncompahgre—"lies a mesa that contains about twenty-two thousand acres. Then down below, at the mouth of the river, is another stretch just twice as large. All that is needful to make that productive farming land is water. A company here is building a canal which will be twenty-seven miles long and will cost a hundred thousand dollars. It takes the water out of the Uncompahgre away up by the Cantonment, leads it along the foot of the wooded bluffs behind the mesa, and can furnish enough to water the whole expanse. If you have a farm there, all you have to do is to select half a mile square or so —there's heaps of it left untouched as yet,—pay $1.25 an acre, dig side ditches and draw as much water as you need at so much an inch rental from the company. That's going to make one vast wheat-field."

"I see, but what next?"

"Well, by the time your wheat is grown there will be mills here to grind it. There is one now at Montrose which will make from seventy-five to one hundred barrels of flour per day, and when the crops get ahead of it other mills will be built. This is not poetry and fancy and talk; it is a settled fact, for the soil has been tried in more places than one, and—but, hello! there's our train!"

Precipitately retreating to our "parlor," we don our dusters and go steaming down toward Grand Junction.

The mountains whence I have just come lift their snow-embroidered heights grandly to the sky, and I can point out nearly all the separate peaks though they are fifty miles away.

> "You should have seen that long hill-range
>   With gaps of brightness riven —
> How through each pass and hollow streamed
>   The purpling lights of heaven —
>
> "Rivers of golden-mist flowing down
>   From far celestial fountains,—
> The great sun flaming through the rifts
>   Beyond the wall of mountains."

On the right, extended a long line of bluffs, close at hand, sprinkled with cedars between which the brick-red soil showed queerly. The strata in the base of these bluffs were yellowish white and had been cut by water into a series of little knolls and spurs like sand-dunes and equally bare of vegetation. They were hot, desolate, and glaring.

The train ran along the edge of the bottom-lands of which these bluffs were the boundary, and on the left stretched a continuous line of farms watered from the river which was hidden in a distant grove of cottonwoods. That the land was rich was shown not only by the flourishing fields of grain, and of Indian corn, but by the luxuriance of sagebrush and greasewood in the uncultivated spaces. This was the Uncompahgre we were following, and at Delta, where the bottom-lands spread out into a spacious plain, we reached its junction with the Gunnison, and passed to its right bank over a long bridge.

Dominating everything here to the northward is that vast plateau, protected from decay by its roofing of lava over the softer substances that make its bulk, which forms the watershed between the Gunnison and the Grand rivers, and is called the Grand Mesa. We know that its surface is hilly and rough, but from here and everywhere else, its edge, as far as can be seen, cuts the sky with a perfectly straight and even line as though it were as level on top as a table. In color it appears dark crimson above the brown and green of mingled forest and exposed rocks that cover its lower front. Looking past it, up the river, we can see the snowy Elks, and a line of rails is surveyed from Crested Butte right down to this point through a series of cañons. There is little opportunity for farming below the mouth of the Uncompahgre, where abrupt walls of red sandstone shut in the river, and sometimes hem it so closely that a road bed had to be blasted out of the cliff. The river has grown, since we saw it last in the Black cañon, to be a hundred yards wide. It still flows deeply and swiftly, but has lost the cataracts. Its color, too, after so much contact with loose earth, has changed from green to turbid yellow. The run along its banks is straight and swift. Generally the track is laid just at the brink, upon the solid rock, and the river is occasionally crossed upon admirable bridges. One of these bridges, I remember, is at a place where enormous cliffs of carmine-tinted sandstone most curiously worn full of little pits and round holes as though moth-eaten, rise sheer from the water to a great height. The strata of these cliffs — which also have bands of yellow — wear away unequally but always in a rounded shape, so that you can see them edgewise, as at a bend, the protuberances take the form of "volutes; and this will continue for long distances unchanged, as if the cliff had been adorned with gigantic beads of molding. It is one of the most interesting stages of the whole journey.

Just east of the Grand are the finest cliffs of all,—great piles of ponderous masonry, fit for the bulwarks of a world, each massive block, a hundred feet or so square, set firmly upon its underlying tier, and the whole rising two or three thousand feet in majestic proportions and colors that please by their softness and harmony. Past these we roll slowly out upon the longest bridge in the state—950 feet—spanning the swift yellow flood of the Grand river just above where the Gunnison enters, and find ourselves at Grand Junction.

XXXI

THE GRAND RIVER VALLEY.

As a true patriot, I should be ashamed to think that Adam in Paradise was more favorably situated, on the whole, than a backwoodsman in this country.
—Thoreau.

A VERY honest little circular—quite a phenomenon among prospectuses—had come into our hands, which gave in terse language the claims that Grand Junction made to the notice of the world and upon the attention of the man who was looking for a place of residence in western Colorado. This honest little circular, toward its end, contains the following paragraph:

We desire, however, to inform all eastern people who may be thinking of coming west, that, while this is one of the most productive valleys in Colorado, it is anything but this in appearance now. Excepting along the banks of the streams or near them, there is probably not a tree to be seen in the valley, unless it was planted since the valley was settled, or within the past eighteen months. The soil has a dull grayish appearance, with hardly a blade of grass growing in it for several miles back from the river, and it produces naturally only sagebrush and greasewood. It is uninviting and desolate looking in the extreme, and yet it is far from being so in reality. We are thus explicit in speaking of the desolate appearance of the country, so that no homesick wanderer in this far-off western land will say when his heart fails him in looking over our valley, that he has been deceived, and that all that has been said of Grand Junction and its surrounding country is a delusion and a snare. If the reader of this lives in the east, he will almost surely be disappointed at first, if he comes out here. It will be the disappointment of ignorance though, for it is only a man who is ignorant of the productiveness of this country who will refuse to believe what is said of it in this respect.

That paragraph put us upon our metal. We were eastern people undoubtedly, but then we had seen "a heap" of Colorado, and the word "ignorance," we would not confess applied in our case. It was therefore with no little curiosity, and something of a resolution to be pleased anyhow (since we had been told we might not be,) that we detached our peripatetic home and slipped into a resting-place upon the customary siding. The glow of the sunset filled the valley with a blaze of yellow light, and the mesas wore chevrons of indigo shadow and pink light to the northward, while the scarred bluffs across the Grand reflected the last rays from burning crests of red sandstone. Weary with travel we threw open our doors, brushed and dusted and bathed,

CASTLE GATE.

while the kitchen was busy, and then sat down to dinner in the cool soft air of the twilight. When it was over a multitude of twinkling lights alone showed where the town lay, and so we left until morning learning more about it.

When we came to the learning, there were persons enough to teach us, besides all the explicit information Mr. William E. Pabor and others have put into type about the new town — the western Denver, the metropolis of—

"Did n't we hear Gunnison called that, too? and Montrose? and—?"

asks the Madame, whose serious mind can never quite become accustomed to local flowers of speech.

Undoubtedly we had; but who shall say which one of them, a century from now, shall not deserve the name? Describe it? That would be merely repetition. Situated, as I have said, in the midst of a level sage-plain, utterly treeless, it is an orderly jumble of brick buildings, frame buildings, log cabins, tents, and vacant spaces. It is South Pueblo or Salida or Durango, or Gunnison of two years ago over again. The more important question to be answered, is, why is a town built here at all? It is here in anticipation of the agricultural productions of the valley by which it is surrounded, water for the irrigation of which is supplied by the largest river in Colorado, and therefore inexhaustible.

A year before the railway came, speculators, chiefly from Ruby and Irwin, who had no dread of loneliness, went to this point and started the town. "They staked off several ranches," says the report, "and located one irrigating ditch and a town site." This town, which they called Granville, is situated across the Grand from the mouth of the Gunnison. A town site was afterwards staked by the Crawford party, and given the name of Grand Junction.

That is the way these marvelously new and flourishing towns are started out here. They reverse the proverb and may be said to be *made* not *born;* or, as Chum puts it, *fititur non nasce.* I couldn't have done that, but it was easy enough for Chum who has been to college; he don't mind a little gymnastics in Latin like that.

In the mountains dividing **Middle park** from **North park** the clustering streamlets pour steadily into Grand lake, whose surface is rarely free from gusts of chilling wind or the shadow of gathering storms. Hidden in heavy forests, it occupies a basin scooped out by the mighty plow of a glacier and held back by moraines and *montonnes* that record a geological history of the utmost interest to the student. About this solitary lake gather gloomy traditions of fierce warfare between Ute and Arapahoe, and since the Indian owners have yielded it to the white men, one of the darkest crimes in the history of the Rockies has happened upon its shores.

From this dark mountain-tarn flows a strong outlet fed by the snows. Its whole youth lies in the depths and gloom of cañons, for range after range open their gates to let it pass, but the gates are narrow and the pathway rough. Thus this river, constantly recruited, more and more the *Grand,* fights its way from the center almost to the western edge of the state. There, when its labor is fairly done, and aid is no longer needed, comes the help of the powerful Gunnison, and doubly strong it rolls westward to the Utah line, and then southwestward till it meets the flood of the Green and both become the Rio Colorado.

Where the Gunnison now empties into the larger stream was once a wide lake embanked by the abrupt and lofty bluffs that now bound the

IRRIGATION.

plain, and whose mesa-top indicates the ancient level of the whole country, out of which the valleys, cañons and lake-beds were eroded.

Into this old expansion of the rivers, had been poured the freight of soil brought down from the mountain-sides, where the varied rocks were being ground to powder under the feet of glaciers, and swept along by gigantic torrents fed with endless meltings. Hither was carried by the swift waters the mingled dust and pebbles of primeval granite, volcanic overflows and sedimentary sands, lime, and argillaceous rock. It was the latest mixture of all that before this had been handled again and again through the fires that upheaved the inner ranges, and the waters that laid down the rocky tables, leaning against their flanks. Into the river-lakes went all this mixture to sink into mud upon the bottom of the quiet sea,—a union of the best elements in all the composition of the western slope of the Rockies. In the whole world you could not find a soil made after a better recipe. Slow changes in the climate proceeded, and the lake drained away and left a valley twenty-five miles long and half as wide, waiting to nourish the farmer's grain and the children's flowers.

The first requisite to adapt it to human service, however, was that the valley should be watered. Thousands of acres of good land in the Rocky mountains from Kootenai to Chihuahua remain worthless because there is not enough water available to spread over them, but at Grand Junction there is no such deficiency. The great drainage of the Grand would not miss all the water that could possibly be used. Already along its margin miles of ranches have been begun, by men digging small and temporary ditches bringing water to irrigate a single farm or a small group of fields in the bottom. These were the first comers who had choice of the whole area. Later two or three larger ditches were made having a greater scope, and now there has just been finished a waterway, led for twenty-five miles along the benches at the base of the Roan or Little Book Cliffs, bounding the plain on the north, which will bring under cultivation thirty thousand acres of valley heretofore unwatered, and may be extended when the population demands. This ditch comes out twelve miles above town. It is fifty feet wide across the top, and is thirty-five feet at the bottom; the depth is five feet, and it delivers seven hundred cubic feet of water each second, at a speed of two miles an hour, though there is only twenty-two inches of slope in each mile of length. A ditch like this costs \$200,000, yet dividends are confidently expected. If anybody can invent a steamer which will not wash the banks, pleasure yachts and freight barges will be put upon it, for it is of considerably greater dimensions than the Erie Canal when first opened. There is no lack of water, therefore. Competent observers say the supply is sufficient for half a million acres, so that the intricate and expensive lawsuits vexing the farmers of the eastern slope can hardly arise here. This abundance is a matter of vital importance, and an inestimable advantage. Water has a value above that of land everywhere in Colorado. Where land, in the valley of the Cache la Poudre, is valued at ten dollars per acre, a water right carries a cash valuation of fifteen dollars per acre and is more easily disposed of. The blessing attending the cultivation of the soil where the water-supply exceeds the area of land, can only be appreciated by those who have seen their crops wither for want of it.

IN SPANISH FORK CAÑON.

It is only recently that this water-supply has become available, however, through the medium of the canals, for any extended farming. Large crops, therefore, cannot be expected until next year, but enough has been learned to make it sure that when the peculiarities of this adobe soil and the looser mesa soil are understood, so that the farmers may know exactly how to supply their irrigation to the best advantage, the most plentiful crops of all the cereals can be produced. We were told that the experiments right here at Grand Junction already, had yielded corn-stalks eleven feet seven inches high; a bunch of wheat having seventy-four stalks in one stool; barley with seventy-six stalks in a stool; oats five and a half feet high; Egyptian millet, one hundred and five stalks from a single seed, weighing thirty-six pounds; four cuttings of alfalfa; Irish potatoes weighing from two to four pounds apiece; cabbages from five to twenty-three pounds apiece; beets, carrots, parsnips, and all the vegetables of equally prodigious dimensions. There can be no question of the extraordinary productivity of this region, and that its agricultural future is to be a very prosperous one.

Equally large expectations are held at Grand Junction, Delta, Montrose on the North Fork, and in all the adjacent lowlands, that this whole region will prove a great fruit-bearing country. The plentitude and excellence of the wild fruits along the streams and in the foothills is remarkable, and formed one of the attractions of the reservation in the eyes of the Indian. The similarity in soil, climate and altitude to the fruit-growing region of Utah is adduced, and, in respect to grapes, peaches, plums, apricots, nectarines, and all the small fruits, successful experiments have justified all the arguments. Just below Ouray, last year, a ranchman raised seven thousand quarts of strawberries for market. I saw watermelons and muskmelons growing finely on Surface creek at the foot of the Grand Mesa near Delta, and everywhere you will find young fruit trees doing well and uninjured by winter, which is always mild so far as known, the thermometer rarely indicating cold below zero, and the snowfall in the valleys being light. "This new Colorado has a climate essentially different from that of old Colorado and the country east of the Continental Divide. It is the climate of the Pacific coast modified only by altitude and latitude. The air currents come up over the valleys, plateaus, hills and mountain sides, fresh from the ocean currents that wash the Pacific shores. These ocean streams are heated under an equatorial sun and sweeping north around the circle of the earth, temper the whole western slope."

In this neighborhood, too, are splendid grazing regions, which are rapidly filling with cattle. I have before me a trustworthy scrap from the *Colorado Farmer* on this point: "The face of the country," says the writer, referring to the hill regions of the Uncompahgre and the grand plateaus, "is gently sloping but cut by gulches, ravines and cañons; grass grows luxuriantly from the creek and river bottom to the very tops of the highest plateaus; on the higher uplands there are plenty of pine and piñon trees, in many places interspersed with cedars and aspens; many small brooks and springs course their way down the hillsides; natural shelter is found in every neighborhood from storms when they come, which is seldom. Game abounds in the greatest plenty and taken all in all, it is probably the finest stock range in the state. The quality of the grass is excellent and cures completely. It is different from the plains grass, grows tall with an abundant wealth of leaf, stem and seed. This country is to be the great cattle country of the state. The Rio Grande railway runs through it and will carry the fattened beeves to the market of the mountains and to Denver and even start them on their way to the great markets of the East. There are already cattle there and they are but forerunners of the thousands yet to come."

The important point is, that the wide mountain areas insure summer pasturage without driving to great distances; while the valleys afford good winter grazing. I have been in every cattle region in the United States, and I never saw anywhere such magnificent grass as I have ridden through for miles and miles along the upper part of Surface creek in Delta county. When the herds have so increased that the winter pasturage falls short—some years must elapse before that— the valley lands will furnish an abundance of millet, oats, alfalfa and other grasses, by means of the inexhaustible supply of water which is possible for irrigation.

As further aids to her progress, Grand Junction has easy access to coal, both hard and soft; has limestone in great abundance, and excellent white sandstone for building purposes; while the soil is adapted for making sun-dried adobes or for being made into burned brick, of which material most of the buildings and many of the sidewalks in town are now constructed. Game is common in the neighboring mountains, especially throughout the great wilderness which stretches northwest, and the rivers abound in edible fishes.

At length there comes a day when we are ready to leave Grand Junction and "go West." It is a long ride that lies ahead and we turn our parlor car into a sleeper by setting up the cots and curtains that have not been needed for several weeks. It is late in the afternoon, and when the morrow's light dawns we shall be out of Colorado and among the lakes and deserts, the mountains and Mormons of Utah.

XXXII

GREEN RIVER.

And then the moon like a goddess came
  Over the mountains far,
Wrapping her mantle of silver light
  Over each golden star;
And the cliffs grew grand in the dazzling light,
High as the skies, and still and white.
—FANNIE I. SHERRICK.

HE sweet clear twilight was fading from the cliffs, and had long since left the valley, when it came time to leave Grand Junction. The rising moon beckoned us on, however, and we look forward with eagerness to our journey, for to-night we are to cross "the desert," to span the cañon-begirt current of Green river, and beheld the mountains of Utah. Doubtless the silent hours of the dog watch would finally close our eyelids; but now we bade Bert be sure that the lamps in the parlor-car were well filled and trimmed, for none of us would confess the least desire for sleep.

In a short time the valley of Grand Junction had been left behind, and we quickly passed through the gravelly, grass-covered hills that lie between the river and the cliffs in this region. It was not quite dark, therefore, when all this had disappeared, and our train ran in a swift straight course across an open and level, though by no means smooth plain. Northward it was bounded at a few miles distant by the frowning and banded wall of the Book cliffs, colorless now in the wan light, but distinct in their majestic outline; southward it stretched to the horizon, save where it was broken by the splendid file of the Sierra La Sal—an isolated group of eruptive mountains singularly graceful in contours. The surface of the ground was drab or blue or yellow in color, nowhere quite flat, but divided into low, rounded ridges and conical mounds, by the shallow dry channels, down which have coursed the waters of the powerful storms that at long intervals burst over the desert. Stimulated by the occasional moisture in these channels, a few spears of grass and twigs of wormwood are thrust up through the soil, along their depressions; but between — over the general face of the country,— not a sign of water, vegetation, or animal life appears. It is the repose of utter silence and quietude, a netherworld only half lighted by the worn-out moon. Yet it has a fearful beauty, found in the magnitude of the space—the grandeur of the huge rocky masses faintly but continuously outlined against the bright sky north of us—the wide realms of gray darkness southward—the marvelous brilliance of the moon—the luminous glory of the overspreading dome, unbroken from horizon to horizon, almost as at sea, and so seeming really a part of the globe and not an external thing. These things impress us greatly and emphasize the sense of loneliness and remoteness. No other railway journey in the country, I believe, could reproduce as this does the impressions of an ocean-voyage.

At Grand Junction we leave the Grand river, though our course for some miles is parallel with it and not far remote. Skirting the edge of the great Uncompahgre plateau which lies between the Uncompahgre and Gunnison rivers and the Rio Dolores, the river flows west and southwest through deep gorges in the Jurassic and Triassic rocks as far as the mouth of the Dolores. This river comes in from the southeast, taking its origin in the Sierra de La Plata, and running a most picturesque course. Through its mouth it is supposed the Gunnison, before it was deflected toward its more northern outlet by the slow upheaval of the plateau, once flowed by the way of a cañon which connects the present valleys of the two rivers. This deserted cañon was called by the Utes Unaweep (Red Rock), describing the scenery it presents. The granite rises vertically from the bottom of the valley, in narrow, bas-relief columns, for some hundreds of feet; above, the beds of red sandstone cap it in broken precipices. In some places massive promontories of the granite, whose slow elevation has raised the whole breadth of the plateau upon its shoulders, juts out into the valley worn down through it. The scenery reminds one strongly of the Yosemite.

In the acute angle between the Rio Dolores and the southward bending Grand lies the Sierra La Sal,—a center of drainage in all directions. It is a mass of volcanic rocks thrust up from beneath. Like the Henry mountains, the Sierra Abajo and other groups of that region, these peaks were once covered by a great thickness of sedimentary strata bent over them; but they have been cleaned away, leaving the hard core of porphyritic rock exposed. The original shape of the upthrust was probably that of a huge dome, but the tooth of time has gnawed it into a score or more of clustered mountains rising eight and nine thousand feet above the level of the adjacent rivers. Yet there is no doubt that the summits of mountains like these, as I remarked of the elevations about Abiquiu in New Mexico, mark the depressions in the primitive surface before this prodigious work of erosion and corrosion had begun. One of the streams flows with strong brine, suggesting the name Salt mountains to the group; but the rest give pure, sweet currents when they flow at all, which with many of them is only for a few hours following a storm. The source of Salt creek is in Sinbad's valley,—a steep-walled nook in the mountain-side abounding in crystallized salt.

THE RIVER THROUGH THE CAÑON.

After receiving the Dolores the Grand river flows straight southwest to its junction with the Green, burying itself at first in a deep, narrow, winding cañon in the red beds, then emerging into a valley of erosion surrounded by tremendous cliffs of deep red sandstone, 1,600 to 2,500 feet high, carved in fantastic forms. It rose 8,150 feet above the sea, 350 miles away; it has fallen to 3,900 feet, or an average of more than ten feet in every mile, and delivers to the Rio Colorado about 5,000 cubic feet of water every second. Considering this weight and speed we need not wonder at the profound cañons it has cut, and is still chiseling deeper and deeper, nearly keeping pace with the slow elevation of the land.

The line of ragged, roan-tinted, book-edged cliffs on the north, behind whose battlements stretched an invisible plateau of broken wilderness, covered with grass, but almost treeless and waterless, where the traveler must not leave the Indian trails,—this line of massive and

TRAMWAY IN LITTLE COTTONWOOD CAÑON.

vari-colored cliffs stretched all the way to Green river (and far beyond it,) rising there into the loftier and bluer bluffs which have been named Azure, and, in the sunlight, seemed carved from cobalt. Between their towering portals, through the corridors of Gray cañon, came the yellow flood of the Green river, sweeping with enormous power from north to south, and crossed by us toward midnight upon a long and lofty bridge. We looked down with eager eyes upon its swift flood of chocolate-colored water, half as broad as the Missouri—twice as deep and impetuous. We wished it had been daylight, that the pregnant mysteries of the half-darkness might be revealed, wherein distant forms full of curious interest were dimly suggested. They told us that here, at noonday, the passenger upon the railway can see the summits of the broken walls that form the Grand cañons of the Colorado, fifty miles to the south.

But all the "grand cañons" are not away in the southern drylands. The whole track of Green river from its birth to its death runs in gorges whose depth and splendor excite our amazement. There are few rivers in the world that have a history so striking; and if, as is fair, we count it one stream from the Wind River mountains to the Pacific, the mighty river is without a peer in its erosive work.

Its source is at the southwestern corner of Yellowstone park, in Wyoming; its mouth, two thousand miles southward, at the head of the Gulf of California. The present writer pens with gratification the record that he has seen both these points. Its upper course lies in open, or wooded valleys, where sparkling, trout-haunted rapids alternate with pools in whose mirror-smooth surface the images of fleecy clouds play with the tremulous forms of snowy peaks. Then it learns lessons for the hard-working future among the plains and buttes of southern Wyoming, cutting through its first obstacle where the Alcove bluffs rear their gaudy crests abreast of Bitter creek.

Here is a little village, settled long ago by emigrants and cattle-breeders, and here, in 1869, Major J. W. Powell, now Director of the United States Geological Survey, and chief of the Bureau of Ethnology, began his celebrated exploration of the river in small boats, which ultimately navigated all the thousand miles of almost continuous cañons that lay unexplored, uncanny and perilous before them. Wonderful stories of it were believed by the frontiersmen. Boats, they told Major Powell, had been carried into overwhelming whirlpools, or had been sucked with fearful velocity underground, never to reappear, for the river was lost in subterranean channels for hundreds of miles. Falls were reported, whose roaring music could be heard on distant mountain-tops; and the walls were so steep in the desert, that persons wandering on the brink had died of thirst, vainly endeavoring to reach the waters they could see below. The Indians believed the river had been rolled into an old trail that once led from their hard home to the beautiful balmy land of the Hereafter in the great west, in order to keep them away until death gave their release.

Undeterred by these tales, the explorers started. Their story has been told by Major Powell himself in his report to the government, and in magazine articles. Before him Macomb, Ives, and Newburry had seen the southern gorges; since then Dutton, Homes and others of the Geological Surveys have surveyed, mapped and sketched the strange scenery of that strange river. Yet to no one can anything but seeing with his own eyes bring more than the faintest conception of the reality. And here we are, at midnight, in the very midst of it—northward and southward lie the profound chasms, the immeasurable and uncountable cliffs;—under our feet flows the mighty river that carved them out and connects them into one.

What a voyage was that of Powell's! The fantastic architecture of the Alcove foothills, with the gleaming points of the Uinta range in the south; the ever-changing panorama of the badlands — scenery for Hades; the vermillion gateway opened through the snow-capped mountains, called Flaming Gorge, where lies a vast amphi-theatre, each step built of naked red sandstone, and a glacis clothed with verdure! Then the cautious advance, after letting the unladen boats down with ropes over foaming rapids; threading gorge and cañon and flume, each characteristic in some new way, and separated by little parks and lowlands filled with trees and quaintly shaped rocks from the next; always hemmed in by lofty and brilliant walls; on to the Cañon of Lodore and Ashley's Falls where years ago a party of men were drowned and where Powell loses one of his boats. "Just before us," he says at one point, "the cañon divides, a little stream coming down on the right, and another on the left, and we can look away up either of these cañons, through an ascending vista, to cliffs and crags and towers, a mile back, and two thousand feet over head. To the right a dozen gleaming cascades are seen. Pines and firs stand on the rocks, and aspens overhang the brooks. The rocks below are red and brown, set in deep shadows, but above they are buff and vermillion, and stand in the sunshine. The light above, made more brilliant by the bright-tinted rocks, and the shadows below more gloomy by the somber hues of the brown walls, increase the apparent depth of the cañons. . . . Never before have I received such an impression of the vast heights of these cañon walls; not even at the Cliff of the Harp, where the very heavens seemed to rest on their summits."

Below the Cañon of Lodore was found the wonderful Echo Rock, where the Yampa enters; next the Whirlpool, where the boats waltz down the tortuous and bowlder-strewn rapids in a merry dance of eddies over which the oars have no control. "What a headlong ride it is! Shooting past rocks and islands! I am soon filled with an exhilaration only experienced before in riding a fleet horse over the outstretched prairie." Passing through the "broad, flaring, brilliant gateway" of Split mountain, and down a series of rapids in a more open region, the mouths of the White and Uinta rivers are passed, and the river brings them to the chaotic scenery of the Cañon of Desolation.

This cañon is very tortuous, and many lateral cañons enter on either side. The great plateau, in which they are sunken, extends across the river east and west from the foot of the Colorado Rockies to the base of the Wasatch. It is eight thousand feet above the sea, and therefore in a region of moisture, as is attested by the forests and grassy vales. On these high table lands elk and deer abound, and they are favorite hunting-grounds for the Utes, whose trails cross them. Nothing of this, however, is seen from the river level, where the voyager is surrounded by a wilderness of gray and brown crags. "In several places," says Powell, "these lateral cañons are only separated from each other by narrow walls, often hundreds of feet high, but so narrow in places, that where softer rocks are found below, they have crumbled away and left holes in the wall, forming passages from one cañon into another. . . . Piles of broken rock lie against these walls; crags and tower-shaped peaks are seen everywhere; and away above them long lines of broken cliffs, and above and beyond the cliffs are pine forests. . . A few dwarf bushes are seen here and there, and cedars grow from the crevices—not like the cedars of a land refreshed with rains, great cones bedecked with spray, but ugly clumps, like war clubs beset with spines."

Various adventures carry the plucky party through and beyond this gorge down to where our railway bridge spans the river with its tenacious links. They note the existence of an Indian ferry of rude log-rafts somewhere near here, but there was nothing to induce their stoppage for more than a night. Now, those of us who are minded some day to behold the wild crags of Desolation cañon will reverse Major Powell's course, and embarking at this railway station on the river bank go up the Green, through the Azure Cliffs and fifty miles beyond. Or, turning southward, our boat may equip itself for a longer journey, and our minds make ready for even more marvelous and memorable sights, in the profundities of the cañons of the Rio Colorado, below the junction of the Green and the Grand. If so long a journey is forbidden, there is much delight, with the advantage of easy and safe navigation, to be found between the railway and the mouth of Grand river.

A few miles after leaving the railway, downward bound, the voyager would get among curious bluffs and buttes that would interest him all the way to the mouth of the San Rafael, a strong tributary from the west, up which passed one of the principal overland trails from New Mexico to Utah. If he is interested in archæology, Indian "relics" in abundance will reward his search along the banks. The river is tortuous here, but deep and quiet. Sometimes there is a narrow flood-plain between the river and the wall on one side or the other, the peninsulas being pleasantly wooded. The walls are orange-colored sandstone, and vertical, but not very high. At one point, where the river sweeps around a curve under a cliff, a vast hollow dome may be seen, with many caves and deep alcoves. The doublings of the river are many; one loop carries you nine miles around, yet makes only six hundred yards of headway. Gradually the chasm of the river grows deeper; the walls are systematically curved and grandly arched; of beautiful color, and reflected in the quiet waters with deceiving distinctness.

This is Labyrinth cañon, or, as the Indians called it, *Toom'-pin wu-neár*,—the Land of Standing Rocks. "The stream is still quiet, and we glide along through a strange, wierd, grand region. The landscape everywhere, away from the river, is of rock—cliffs of rock; tables of rock; plateaus of rock; terraces of rock; crags of rock—ten thousand strangely carved forms. Rocks everywhere and no vegetation; no soil; no sand. In long gentle curves the river winds about these rocks. When speaking of these we must not conceive of piles of bowlders, or heaps of fragments, but a whole land of naked rock, with giant forms carved on it: cathedral-shaped buttes, towering hundreds or thousands of feet; cliffs that cannot be scaled, and cañon-walls that shrink the river into insignificance, with vast hollow domes, and tall pinnacles, and shafts set on the verge overhead, and all highly colored—buff, gray, red, brown and chocolate; never lichened; never moss-covered; but bare, and often polished."

Below the Labyrinth is Stillwater cañon, forty miles long. Its walls at the lower end are beautifully curved, as the river sweeps in its meandering course. Suddenly gathering swiftness it rushes hastily forward to unite with the current of the Grand. These streams join their floods in solemn depths, more than twelve hundred feet below the general surface of the country. Up the Grand you look into another "labyrinth." It is the central artery toward which innumerable side-cañons concentrate. In every direction they ramify, deep, dark and impassable to everything but the winged bird. Through such underground passages are sent the waters from the distant highlands, and their confluence fills the whole chasm of the Grand with a turbid stream.

Climb out, laboriously and cautiously, ascend one of the fantastically-formed buttes that rise above the level of the plateau, and what a world of grandeur is spread before the eye! Nothing one can say will give an adequate idea of the singular and surprising landscape,—nothing in art or nature offers a parallel. Below lies the cañon through which the Colorado begins its wonderful course. It can be traced for miles, and occasional glimpses of the river caught. From the northwest comes the Green, in a narrow, winding gorge; from the northeast the Grand, hidden in a cañon that seems bottomless. "Away to the west are lines of cliffs and ledges of rock,—not such ledges as you may have seen where the quarryman splits his blocks, but ledges from which the gods might quarry mountains, that, rolled out on the plain below, would stand a lofty range; and not such cliffs as you may have seen where the swallow builds its nest, but cliffs where the soaring eagle is lost to view ere he reaches the summit. . . . Away to the east a group of eruptive mountains are seen—the Sierra La Sal. Their slopes are covered with pines, and deep gulches are flanked with great crags, and snow fields are seen near the summits. So the mountains are in uniform—green, gray and silver. Wherever we look there is but a wilderness of rocks; deep gorges where the rivers are lost below cliffs and towers and pinnacles; and ten thousand strangely carved forms in every direction; and beyond them, mountains blending with the clouds."

I cannot go on to tell of the profound crevices in the crust of the globe beyond, where the Rio Colorado, taking its name from its vermillioned borders, flows a full mile below the surface. A whole volume like this would not suffice to portray fully the pictures and the teaching of a single day's ride down that engulfed stream, or an hour's march along the giddy brink. Only one man, Captain C. E. Dutton, has ever given anything approaching an adequate description. He lived on the plateau and studied it for years; and he tells us that it is a long time before the unaccustomed mind can come to have any real comprehension of the magnitude and the sublimity and the exquisite beauty of what the cañons above and below have to show to the attentive eye. Nothing in the wide world equals or compares with it in its peculiar and amazing beauty. and force.

But the fanes and museums of these rock-gods are guarded against the too easy profanation of human curiosity. The terrors of personal discomfort and danger surround them. Enduring and brave must be the horseman or canoeist—what a trip for the Rob Roys of the future!—who penetrates this naked wilderness and feasts his eyes on the riches of novel color and form spread before him in the glory of the setting sun!

The dog watch comes. The gray waste of sterile land sweeps steadily past. The stars wheel slowly along their cosmical paths. Utter loneliness envelopes us as we rush forward with direct and tireless speed. The rolling music of our progress, and the solemnity of our thoughts as we ponder what we have seen and heard, quiet mind and body, the lamps are turned down, the curtains drawn, and silence and darkness reigns in our car, as over the night-beleaguered desert save where some official passes silently through, shading his lantern with his hand.

SALT LAKE CITY.

XXXIII

CROSSING THE WASATCH.

> The splendor falls on castle walls
> And snowy summits, old in story;
> The long light shakes across the lakes
> And the wild cataracts leap in glory.
>   Blow bugle, blow; set the wild echoes flying;
>   Answer, echoes, answer, dying, dying, dying.
>   —TENNYSON.

ITH the first full light of dawn, on the morning after leaving Grand Junction, the vigilant Madame was awake, and we heard her calling upon us from her curtained corner to wake up and look out of the window. Well, as the Shaughran said when punished for his foxhunt on the Squire's horse, "It was worth it," even at the expense of the morning nap. Here was something different from anything seen before.

We were far inside the boundary of Utah Territory and were already beginning to climb the first steps toward the heights of the Wasatch—the western bulwark of the Rocky mountains. The way lay up the South Fork of the Price river, along a broad valley sunken between enormously high walls of sedimentary rock whose horizontal stratification betrayed no signs of disturbance. How long must the waters of the palæozoic sea have surged against the primitive granitic caves and lava-masses—how steady must this part of the earth's crust have remained for ages—to let these thousands of feet of rocky tablets be piled up! And then, when it was done; when the slow upheaval had come, and the water had gradually been drawn off; how patiently did the centuries wait while these great depressed spaces were cut down and the material carried away to be spread—who knows where?

Here mountain-like table-lands stretched their white and cedar-spiked terraces, one above another to the plateau-top, for scores of miles out from the range against which they were braced. The water and the sand-blast of the fierce winds had worn their exposed cliff-faces, and sometimes carved their crests (now gold-tipped by the first sunbeams) into fantastic shapes that recalled pictures in the Dakota badlands or the grotesque monuments near Colorado Springs. In some places they were honeycombed with round holes, connecting pits and fissures, like a prodigious display of Arabesque fret-work; elsewhere they would stand massive and plain. As we proceeded colors began to appear,—yellows, warm browns and pale reds, against which, in thorough keeping, grew the bent and aged forms of junipers. In the soft gray of the morning light, nothing could be more pleasing than these worn and variegated battlements, between which for miles and miles the road winds its way. Every stupendous headland was a new rendering of the general idea—a novel design coherent with hundreds of its fellows; and of each the eye was afforded several altered aspects as the train changed its point of view.

Finally we attained a higher level, and the cliffs came nearer, became more precipitous and the inter spaces more green. This was Castle Valley. We had risen and dressed ourselves and were thinking of breakfast. The sun had come high enough over the "great, lone land" in our rear to shoot his beams half way down the projections of the dewy and glittering cliffs, when the train came to a stop, though there was only a side-track. Stepping to the platform to enquire why, we came with all the shock of complete surprise face to face with what to me, is the most inspiring, as a single object, of all the marvelous scenes between the Plains and the Salt Sea. This was Castle Gate.

The cañon here becomes very narrow and tortuous, with picturesque defiles opening here and there and conducting tiny streams, swelled in spring to noisy torrents. Trees and bushes in great abundance grow on the narrow banks of the river and swarm up the rough heaps of rocks that bury the foot of the cliff on each side. Just here, these cliffs are several hundred feet high and exceedingly steep, showing great ledge-fronts as upright and clean of vegetation as the side of a house. All the rocks are bright rust-red, darker and lighter here and there; and over all arches a sky, violet-blue, vivid, and immeasurably deep, for you may look far into it, as into water that lies quiet and luminous under the sunshine.

Now out from this wall on one side pushes a great projection half way across the valley, crowned on its point by a round turret. This is on the left or southern side. Opposite it has been left standing an enormous natural wall—a thin promontory projected from the face of the mountain as Sandy Hook stretches narrow and straight out in the ocean beyond the Atlantic coast-line. From base to combing it rises sheer and toppling whichever way you scan it; and on the western side the topmost ledges overhang. Here the face is scarred not only by the horizontal lines denoting the separate strata, but also by vertical gashes of cleavage some of which are strongly marked cracks extending from top to bottom. These show how, by the continual scaling off of enormous slabs on each side, under the prying levers of heat and cold, moisture and weight, this once thick headland has been reduced to a thinness so contracted that its thickness in proportion to its height is no greater than that of Cleopatra's Needle or any other monumental shaft; while the narrowed peninsular, which connects it with the main crag, has only the proportions of a garden wall; but what a wall! for it is eight hundred feet from its weedy top to the foundation. You can count in the patches of freshly exposed rock on its surface how season by season it is diminishing; and one great crack almost splits its extreme edge in twain so that some day, with an earth-jarring crash, half the thickness of this noble remnant will drop to its base and burst into dust and fragments.

Heedless of such a castatrophe, and unmindful of the grandeur of their home in human eyes, birds build their nests in the crevices and crannies that are nicked into its crimson front, and bats shrink from the light into the seams that make a network upon its sides. Great gaps mar the regularity of its sky line; but these mark the ruinous hand of time and add to the antique grandeur of the pile. We cannot take our eyes from it, and forget the even more lofty walls and pinnacles opposite, which have not the advantage of the isolation, and the Olympian dignity and pose, of this daring pier.

A little group of Indians on horseback, in the full toggery of Uinta Utes, were jogging along the road beside the track when presently we emerged from Castle Valley and drew up at Pleasant Valley Junction. Here a branch road comes from some important coal mines sixteen miles to the westward. This coal occurs in a bed eleven feet in thickness, and so situated as to be worked very conveniently. The mines were opened four years ago, and a railway built to them from Provo — a distance of sixty miles. This was bought by the Denver and Rio Grande and a part of it was utilized. Now all the coal comes down from the mines by gravity, and the locomotive is required only to haul up the empty cars.

This coal is bituminous, and of better quality than that from Rock Springs, in Wyoming, which it is gradually displacing in the Utah markets, since it is found to give more heat, ton for ton. Its introduction was a boon to the people generally, for instead of seven dollars and fifty cents a ton, with occasional extra prices, they now pay only five dollars, and get a better article. The mines are operated by the Pleasant Valley Coal Company, who employ about one hundred men and produce a daily output of three hundred tons, which is constantly increasing to meet the growing demand.

After leaving Pleasant Valley Junction the ascent of the Wasatch was begun in earnest, but, though a long pull, the grade was not remarkably steep, nor was the cañon (worn through a red pudding-stone,) astonishing in any way, while always interesting. By the time breakfast was fairly out of the way, the summit had been reached and the descent began through the cañon of the Spanish Fork into Mormondom.

A few miles down we came to Thistle Station, a place of some consequence because it is the railway outlet for the large San Pete valley, which stretches nearly along the western foot of the Wasatch until it emerges into the valley of the Sevier. This valley is reached from the westward by a narrow-gauge railway line, built years ago by Mormon capital. It enters it from Nephi by the way of Salt creek and terminates at Wales, where there are coal mines worked by Welchmen and operated by English capital. The San Pete valley is not particularly interesting. Yet the little settlements back in the eastern foothills where the many streams come down are pleasant enough.

This valley became famous in 1865-67 as the scene of the San Pete Indian War. "Several companies of the Mormon militia were mustered here, and held the mountains and passes on the east against the Indians, guarded the stock gathered here from the other settlements that had been abandoned, and took part in the fights at Thistle creek, . . . and the rest, where Black Hawk and his flying squadron of Navajos and Piutes showed themselves such plucky men."

Toward the lower end of the valley stands the important town of Manti, its suburbs encroaching on the sagebrush. "As a settlement," says Phil Robinson, the most recent traveller thither, "Manti is pretty, well ordered and prosperous. . . . The abundance of trees, the width of the streets, the perpetual presence of running water, the frequency and size of the orchards, and the general appearance of simple, rustic comfort, impart to Manti all the charateristic charm of the Mormon settlements." Robinson says that the people in that region are chiefly Danes and Scandinavians. "These nationalities contribute more largely than any other—unless Great-Britishers are all called one nation—to the recruiting of Mormonism, and when they reach Utah maintain their individuality more conspicuously than any others." The temple at Manti will be something worth going far to see when it is completed. "The site, originally, was a rugged hill slope, but this has been cut out into three vast semicircular terraces, each of which is faced with a wall of rough hewn stone, seventeen feet in height. Ascending these by wide flights of steps you find yourself on a fourth level, the hill-top, which has been leveled into a spacious plateau, and on this, with its back set against the hill, stands the temple. The style of Mormon architecture, unfortunately, is heavy and unadorned, and in itself, therefore, this massive pile, one hundred and sixty feet in length by ninety wide, and about one hundred high, is not prepossessing. But when it is finished, and the terrace-slopes are turfed, and the spaces planted out with trees, the view will undoubtedly be very fine, and the temple be a building that the Mormons may well be proud of."

The lower part of the valley is undulating,—"for the most part a sterile-looking waste of greasewood, but having an almost continual threat of cultivation running along the center," until it suddenly opens, at Mayfield, into a great meadow of several thousand acres. Passing on to the Sevier, volcanic hills and benches shut in the valley, but the bottoms along the river were level, grassy, "clumped with shrubs and patched with corn-fields." Here there are frequent settlements, one of the most important of which is Salina, where the alkalies that infect the soil of all this region are concentrated in salt-beds that have long been dug for export. The Denver and Rio Grande Company have projected a track from this point due east to join its main line and thus secure not only the salt trade, but tap this farming and grazing region, which some day will be of great consequence, for there is plenty of water. Furthermore, this same company has laid a sort of preëmption right upon a railway route surveyed westward from Salina toward the Pacific coast through southern Nevada and central California, which will pass close by the Yosemite and the Caleveras groves of Big Trees.

We had gathered all this information from books and good friends, and the recollection of former reading, while

"like Iser, rolling rapidly,"

we descended the Spanish Fork. This cañon is not rough and cliff-bound, but its sides, though steep, are rounded, and soft walls of greenery —small bushes, rank grass and tufted groves of aspen and oak; while the river purls along the narrow depression under the continuous shade of young maples, alders, oaks and other shrubbery. A wagon road is to be seen most of the way showing long use. The rude kennels built for temporary use by the railway-makers have not yet had time to crumble, and add a picturesque note to the pleasant vale. Here and there were camps of emigrants, or of railway people. That they were Mormons was plain by the comfortable, home-like appearance of each bivouac, where buxom women were tending to the children and carrying on the ordinary duties of housekeeping in houses of cloth and kitchens made under booths of maple boughs. Children and dogs and donkeys abounded, and at two or three camps we caught sight of pet fawns. This cañon was formerly an Indian highway, and through it, in days gone by, more than one incursion of Navajos and Piutes has swept down upon the settlements, "bringing fire and the sword." Through it also came, long, long ago, the pious friars,—explorer-priests —bent upon the conversion of the Indians; and the digging up of a few coins and other relics of their visits has given the name of Spanish Fork to the stream, which really is a "fork" of no other water course, but empties directly into Lake Utah.

As we emerge from the cañon a wide, haze tinted sketch of mountains and green-gray plains, with a touch of steel-white water, greets the eye to the westward, where stretch the arid deserts and volcanic ranges of Utah and Nevada. Southward, nobly tall and rugged, rise the hights of the Wasatch, with the magnificent pyramid of Mt. Nebo, overtopping the rest, exalted above the sunlit clouds and crowned with early snow wreaths. It is the last of the great, lone, Rocky Mountain peaks that we shall see, and a worthy reminder of the splendid scenery in whose presence our life has been expanded and glorified during the bygone months.

XXXIV

BY UTAH LAKES.

> Straight mine eye hath caught new pleasures
> Whilst the landscape round it measures;
> Russet lawns, and fallows gray,
> Where the nibbling flocks do stray;
> Mountains, on whose barren breast
> The laboring clouds do often rest;
> Meadows trim with daisies pied,
> Shallow brooks and rivers wide;
> Towers and battlements it sees,
> Bosomed high in tufted trees,
> Where, perhaps some Beauty lies
> The Cynosure of neighboring eyes.
>
> —MILTON.

EBO does not long remain in sight from our windows; for speedily we swing out into the sloping valley land, bringing into close companionship on the right the northern half of the Wasatch range, which, were it a trifle more arctic and bristling aloft, would remind us strongly of the Sangre de Cristo. On each side now are spread wide areas of grain field and grassland, with abundant hedges and thickets and orchards surrounding clusters of houses and barns. The train makes frequent halts at little villages, which seem to have been built along both sides of the track, because the reverse process occurred and the rails were laid in the middle of the main street. The first of these farmer-settlements was Spanish Fork (the Palmyra of old settlers) where, not long ago, a copper image of the Virgin Mary was dug up, together with some fragments of a human skeleton. "This takes back the Mormon settlement of to-day to the long ago time when Spanish missionaries preached of the Pope to the Piutes, and gave but little satisfaction to either man or beast, for their tonsured scalps were but scanty trophies, and the coyote found their lean bodies but poor picking." A few miles further is Springville, hidden in well-watered trees, where a stream tumbling out of a mysteriously dark cañon just behind the town, turns the wheels of extensive flour mills and woolen factories.

It is with growing and animated interest, that we pass on through miles of fertile farmland and come into plain sight of Utah Lake,—a glassy sheet of water beyond which loom through their mist the vague forms of many angular hills. The water is fresh, and none of the barrenness that has smitten the shores of the Salt sea northward accurses this beautiful lake, with which the Indians strangely enough, associated many evil influences and dark legends. Between us and the shore stretch vast meadows of green prairie grasses and bulrushes, upon which herds of sleek cattle and fine horses were grazing. Except upon the western side, where the hills yield no water, there is a semicircle of villages at the feet of the encompassing hills, with checkered fields

BEE HIVE HOUSE.

of grain and fodder between the embowered clusters of houses and the swampy meadows along shore. Sometimes the meadows and gardens, the squares of wheat and Indian corn come clear down to the shore.

Though most of the houses were of adobe, showing signs of long occupancy in the advanced state of orchard and garden, and the homelike air about them, the pioneer's wagon top makes him a good enough house for several weeks in this dry and genial climate, but he builds something better for the winter. The second season, therefore, will find him living in a small, but tight and warm, cabin of slabs, chinked and roofed with dirt. His stables will be low structures of poles thatched with straw or rushes cut at the border of the lake, and his grain will be stacked out of doors.

A great gap in the Wasatch has been in sight for some time, in which lies the source of the Provo river, and down here upon its banks is Provo—the largest town on Utah Lake. We have "Sinners and Saints" open before us as we draw up at the station; and the Madame reads to us what the author has to say about this very town:

"Visitors have made the American Fork cañon too well known to need more than a reference here, but the Provo cañon, with its romantic waterfalls and varied scenery, is a feature of the Utah valley which may some day be equally familiar to the sight-seeing world. The botanist would find here a field full of surprises, as the vegetation is of exceptional variety and the flowers unusually profuse. Down this cañon tumbles the Provo river; and as soon as it reaches the mouth . . . it is seized upon and carried off to right and left by irrigation channels and ruthlessly distributed over the slopes. And the result is seen, approaching Provo, in magnificent reaches of fertile land and miles of crops. Provo is almost Logan [in Cache Valley] over again, for though it has the advantage over the northern settlement in population, it resembles it in appearance very closely. There is the same abundance of foliage, the same width of water-edged streets, the same variety of wooden and adobe houses, the same absence of crime and drunkenness, the same appearance of solid comfort. It has its mills and its woolen factory, its 'co-op.' and its lumber-yards. There is the same profusion of orchard and garden, the same all pervading presence of cattle and teams. . . . The clear streams, perpetually industrious in their loving care of lowland and meadow and orchard, and so cheery, too, in their perpetual work, are a type of the men and women themselves; the placid cornfields, lying in bright levels about the houses, are not more tranquil than the lives of the people; the tree-crowded orchards and stack-filled yards are eloquent of universal plenty; the cattle loitering in the pasture contented, the foals all running about in the roads, while the wagons which their mothers are drawing stand at the shop door or field gate, strike the new comer as delightfully significant of a simple country life, of mutual confidence, and universal security."

At Springville and again at Provo, the train was surrounded by a flock of little girls who held up to the windows baskets of fruit—apples, pears, raspberries, plums, grapes and peaches. They sought buyers very prettily, offering whole handfuls of the fruit for five cents. Everybody bought it, for nothing could be more welcome after the weary journey. The Madame rushed out to the platform and proceeded to empty the basket of one gentle speculator whose frock was white and clean, but whose shapely legs and feet were bare and brown. She wore no hat, and there fell down her straight young back a heavy braid of beautiful corn-silk hair tied at the end with a bow of cherry ribbon. Her figure and manners were full of *naïve* grace. As the bargain was concluded and we rolled away, the Madame came near kissing her good-bye, and we heard some one humming

"Happy little maiden she,
Happy maid of Arcadie."

But was it this, or another little maid, or both, she had in mind, while the soft light shone in her eyes?

Nephi, the next station—a mass of orchards surrounding straggling streets of doubled-doored gray houses—is memorable because of the remains of fortifications that surround it, with lesser defences near many of the houses. These consisted of thick mud parapets pierced for rifles; and they recall the dangers these pioneers had to encounter from the "Lamanites," as they called the redskins. "Young men tell how as children they used to lie awake at nights to listen as the red men swept, whooping and yelling, through the quiet streets of the little settlement; how the guns stood always ready against the wall, and the windows were barricaded every night with thick pine logs."

The beautiful valley of Utah Lake has now been left behind, and the scene returns to the familiar sagebrush and volcanic scoria, through which a small river of yellow water finds its way, and we follow all its curves. The river is the Jordan, so called because it connects the Utah with the Great Salt Lake, as its namesake does Galilee and the Dead Sea. But the yellow river and its desolate ridges are presently passed, and there opens out on each side a vista of great fields of wheat and tasseling corn; of orchards heavy with ripened fruit, and meadows sere with summer heat; and of houses hidden in trees and hopvines, and touched with the brilliant points of climbing roses and honeysuckles, or the lofty standards of the hollyhock, flying by like the panorama of a dream.

Up the grand slope of the Wasatch beyond, stretches a mass of houses and a forest of shade trees, that are sweeping every instant nearer. Shade of Jehu, how we are tearing along! Swish! That was a smelter. Swish again! That was a furnace. Crash! Bang! Salt Lake City! Shall we halt? No, only a few moments to watch the crowd alight and wrestle with the hotel runners; and also to detach and arrange for the side tracking of our two household cars. We will keep the coach and go on to Ogden while we are "in running order" as Chum says. Then we will come back to-night and stay in Salt Lake City as long as we please. So with a parting admonition to Bert we take our seats and are moving onward once more.

Here again the track for a long distance runs along the middle of a surburban street slowly traversed,—a street of lowly houses, each in its dense garden. It is not at all a bad notion of the whole city, which that glimpse gives the traveller, but shortly it is exchanged for a sage-bush-plain, followed by a region reminding us strongly of the St. Clair flats in Canada. Meadows and marshes, vividly green, stretch to the westward, diversified by planted groves of cottonwood, while mountains rise close at hand on the east. Here and there pools of calm water flit by, on whose surface large flocks of snow-white gulls sit motionless. It is a great place for blackbirds, also,—Brewer's grakle and the yellow-headed blackbird—one of which races with the train, apparently just to show how fast he can fly. Presently the ground becomes dryer and shows wide cultivation. Stacks of hay and straw dot the level and unfenced expanse, but the houses and barns of the farmers are all at our right along the foot of the hills. They are pleasant homes, embowered in orchards, and the whole scene is sunny and peaceful.

The soil is black and loamy, the foothills green and studded with blooming farms and homesteads, the lowlands lush with long grass and willow thickets. Westward, the scene might be a *replica* of, say, the coast of North Carolina, for now the Great Salt Lake is in full view, and the mist which hides the mountainous islands and western shore, leaves its expanse as limitless as that of the open ocean, whence no salter breeze could blow than this morning air. We gradually approach the shore, or its bays bend forward to our straight line, and we leave the fields behind to skirt and cross a great expanse of salt whitened mud flats, where chestnut-backed plovers flit about as the only sign of life. On higher ground, just beyond, a frail pier or landing stage runs far out into the lake, where is moored a small steamboat, and two or three sail-boats rest on the gently ruffled water. This is a bathing resort and picnic grounds, which hereafter will be made more of than at present. Beyond, for miles and miles, the country seems to have been one continuous wheat-field, for the golden stubble stretches in vast unfenced spaces, and we can count dozens of huge yellow stacks that have been reaped. A long ridge of dry gravel is traversed, a vista of valley land, filled full of groves, and orchards, market gardens and neat houses, opens at the base of high rocky walls and the locomotive gives its last long shriek, for this is Ogden, the terminus of our westward jaunt,— 771 miles from Denver, 2,500 miles from New York, 864 miles from San Francisco.

When luncheon was over, I sat me down to my work, and the Madame began putting on her hat, making quite sure that it was straight, nor leaving the neighborhood of the mirror until wholly satisfied on that head.

"That is complimentary to Ogden!" I observe with a rising inflection.

"Not particularly," she answers slowly. "I would want my hat to sit straight and my feather be right if I were going into a camp of Digger Shoshones. It wouldn't feel right otherwise."

I do not argue the question. Turning to Chum she enquires sweetly (ignoring anybody else) if he will go with her on a stroll of exploration.

That young man is just filling his pipe, and the expression of anticipated delight fades utterly from his countenance.

"E-r-r," he stammers, thinking how he may escape. "Thanks, thanks, but I can't very well—I—I have letters to write, you know."

"Yes, I know, everybody has, under certain circumstances. However, I can go alone. *Au revoir!*"

Then I sit down at work. Chum lights his pipe and lazily scratches a postal card to keep up appearances, and silence reigns for an hour or so.

It is put at end by our lady's return.

"Well, what did you see?" we both ask.

"Oh, Ogden is a big collection of little houses and behind each house is a pretty little farm and market garden. There is a ledge beyond the main part of the town, and up there are situated the better houses of the city, with larger gardens and lawns, from which you can look off over the wide plain with bluffs and ridges, in the foreground, catch a glimpse of the lake in the middle distance, and a vision of sharp-pointed mountains on the horizon."

Ogden holds interest at present, chiefly—and it always will I fancy—as a center for transportation lines. Here, in 1869, was welded into a continuous whole the first line of rails connecting the Atlantic with the Pacific. This is the scene of Bret Harte's stirring poem, which tells us

> '. . . what the engines said,
> Pilots touching—head to head, .
> Facing on a single track,
> Half a continent at each back."

That event was an occasion of public rejoicing, and the success of those lines at that time was a matter of public concern.

Since her one line east and west was first connected, Ogden has seen a great growth in railways. The traveller may now go northward into the mining regions of southern Idaho, or on to the quartz and placers and the silver ledges of Montana; or, still further, around Pend' Oreille and Cœur d' Alêne and down the majestic Columbia to Oregon, Washington Territory and British Columbia; he may go westward to California and the Pacific; he may go southward to the farms and mines of southern Utah; or eastward into the heart of the Rockies and so through to the Atlantic over the route we ourselves have just passed.

Ogden has some thousands of people claiming it as home; and besides the large patronage of the railways it is the supplying-center, and the market for a considerable farming district in southern Idaho. It is a busy and enterprising and growing town. Its union station is a sort of narrows through which the larger part of all exchange of men and goods between the east and west must drain; and there is excitement and variety enough to keep alive the attention of the dullest witness. The train that brought us in the morning had sent a merry crowd on to San Francisco—a train-load of acquaintances in jolliest mood, for the other incoming trains contributed very few to the company bound westward. Now the arriving train of the Central Pacific poured across the busy platform another just such a merry company, filling the cars of the Denver and Rio Grande, to which our special was attached, and losing few to the older line.

At Salt Lake City, that evening, our faithful Bert had a good dinner ready for us almost the moment we returned; and, restored to the comforts of our own bed and board, we made an auspicious and good-natured entry to Zion, and to Deseret, the chief city of the Latter Day Saints.

XXXV

SALT LAKE CITY.

> "I have described in my time many cities, both of the east and west; but the City of the Saints puzzles me. It is the young rival of Mecca, the Zion of the Mormons, the Latter-day Jerusalem. It is also the City of the Honey Bee, 'Deseret,' and the City of the Sunflower — an encampment as of pastoral tribes, the tented capital of some Hyksos, 'Shepherd Kings'—the rural seat of a modern patriarchal democracy; the place of the tabernacle of an ancient prophet-ruled Theocracy."
>
> —PHIL ROBINSON.

IT was on a Saturday night that we returned to Salt Lake City. It followed, therefore, as a matter of course, that the ensuing morning was Sunday. Had the calender not been our authority we might have known it from the solemn stillness that prevailed — a contrast very vivid and suggestive after our experience of the Holy Day in the mountain mining towns.

Everybody was eager of course to go to the Tabernacle.

The Tabernacle stands inside the big wall surrounding the "Temple block," and could have been found by simply falling in with any one of the currents of Sunday-dressed people which set toward it from every direction. We went early so as to look about us at leisure.

This square of ten acres was set apart for temple purposes at the founding of the city, and there the pioneers held their first worship. There was built the building known as the Endowment House, and there, thirty years ago, were laid the foundations of the temple, wherein (it is promised) Jesus Christ shall appear bodily to the faithful as soon as it is completed. Reared to a height of eighty feet above the ground, but not yet ready for the roof, its snowy walls gleam in the sun, hot and dazzling.

There is a little time before the services in the Tabernacle and we go over to the new building, picking our way among redoubts of the sparkling blocks of granite. A picture of the building as it will appear when the work is finished, hung under glass, at the closed door of the superintendent's office, and enabled us to get a very good idea of how the great structure would look. The Madame joined the rest of us in admiration for the massive character of every part of the work.

We found that above the enormous foundations the wall had a thickness of nine feet, which decreases to seven at the height of the roof. Nor was this wall hollow, or filled or backed with brick or anything

MORMON TEMPLE, TABERNACLE AND ASSEMBLY HALL.

else, but was made solid throughout of hewn granite. Even the pillars, the partitions, the stairways, the floors and ceilings in many apartments, were of solid matched stone. The beveled window openings through these thick walls are like embrasures of a fort; and the many small rooms in which it is to be divided, will cause the structure to seem more like a prison than a religious temple. In fact it is not designed as a house of worship,—the Tabernacle remains for that—but is intended as a sacred edifice within which various ordinances of the Church, closely allied to those of Masonry, now performed in the Endowment House, shall be celebrated.

The external ornamentation of the great building is original and symbolical in its plan. The wall is pierced with four tiers of large windows, the second and fourth tiers being circular. The keystone of each of the arches over these windows, as well as over the various doors, bears a star in high relief; and between the windows room is found for three tiers of circular bosses, eight on each side, upon which symbols will be carved in high relief. The lowermost of these rows will bear maps of various parts of the world; the second tier, eight phases of the moon; and the topmost tier eight blazing suns. The suns, moons and stars are already cut, but the maps of the earth remain to be carved.

The cost of the temple has been the subject of much public questioning and careless slander. A man assured us that it had cost sixteen millions of dollars. This is certainly an exaggeration. I have the word of President Taylor that the total cost up to the present time has been about two millions of dollars, derived from the church tithings. The same work could be duplicated now for a far less sum; but a large part of this was done before the railway was built, when the stone had to be hauled from the distant quarries by ox-teams. It is supposed that another million dollars and two years more time, will complete and furnish the building. That will be a great day for Salt Lake.

By the time we had finished our inspection of the temple, the Tabernacle had been pretty well filled by the crowds of people who poured into its many doors. One of these streams we followed.

The Tabernacle has been so often described and figured that I need spend little time over it. Imagine an elliptical dome, shingled, set upon a circle of stone piers about twenty feet in height, and you will have an image of this extraordinary building. Were it set upon an eminence it would be as grand in its place (perfectly fitting Utah scenery in its severely simple outlines) as the Acropolis at Athens.

Service in the Tabernacle is held on Sundays at two o'clock in the afternoon. The people assemble not only from the city but from all the country around. Women and children are in great force. The great amphitheatre supplies seats for thirteen thousand people, and it is nearly filled every Sunday. A broad gallery closes around at the front end where the choir sit in two wings, facing each other—the men on one side and the young women on the other. The space between is filled by the splendid organ (back high up against the wall) and by three long, crimson-cushioned pulpit-desks, in each of which twenty speakers or so can sit at once, each rank overlooking the heads of the one beneath. The highest of these belongs to the president and his two counselors; the second to the twelve apostles, and the lowest to the bishops. The acoustic properties of the building are wonderful; a person standing in a certain space near one end, can hear the gentlest whisper, or, that universal test, a pinfall, from quite the other end. A former deficiency of light, has been overcome by the use of gas and electricity; and the chilling barrenness of the vast whitewashed and unbroken vault is relieved by a liberal hanging of evergreen festoons, and trailing wreaths of flowers made of colored tissue paper. These trimmings are far enough away from the eye, and in masses of sufficient size to make their effect very satisfactory.

Every Sunday the sacrament is administered, the tables loaded with the baskets of bread and silver tankards of water (never wine) occupying a dais at the foot of the pulpits, upon which several bishops take their places, and break the bread into fragments. Precisely at two o'clock the great organ sends forth its melodious invocation, and the subdued noise of neighborly gossip, which, as the Madame said, "seemed the veritable humming of the honey bees of Deseret in their house hive," is wholly hushed. The music at the Tabernacle is far-famed in the west, and gives constant delight to all the people. The singing is followed by a long prayer by some one of the dignitaries in or about the pulpits, during which the time is utilized to prepare the bread and water; and as soon as the prayer has ceased a large number of brethren begin to pass the sacred food and drink. Everybody, old and young, partakes, and it is an hour and a half before the communion is completed. Meanwhile some one of the highest officers of the church, or perhaps two or three of them in succession, has been preaching; so that two long hours are exhausted before dismissal. Such was the experience of our visit, and it was an average occasion.

The history of Salt Lake City is the history of the "Mormons,"—of the Church of the Latter-Day Saints in Utah. It begins on the 24th of July, 1847, when Brigham Young, leading the people, who likened their pilgrimage to a second exodus of Israel, emerged from the long cañon that had let them through the westernmost range of the Rockies. As the head of the weary train passed the last barrier they saw spread before their eager vision a huge basin—miles of sage-green, velvety slopes, sweeping down on every side from the bristling mountain-rim to the azure surface of the Salt sea set in the center.

Here, their leader told them, the Lord commanded a halt; here his tabernacle should be raised. It was done, and to-day a populous city stands on the site of the first camp of the religious host,—a city as baffling to describe in its appearance, in its social aspects, in its pervading sentiment as any which can be found in Christendom. It was with an intense sympathy for Mr. Robinson, that I listened to the Madame reading the opening paragraph of the fifth chapter of his "Sinners and Saints," a part of which I have selected as the motto of the present chapter.

Yes, like Mr. Robinson, I would have liked "to shirk this part of my experiences altogether," but the reader would never have pardoned me. "What? Leave out Salt Lake City!" I hear you exclaim. "What's the good of mentioning Utah at all, if you do that?"

Well, to begin with, the city is not on the lake nor within a score of miles of it. When the pioneers came they descended to the foot of the last "bench" in which the foothills yield their rights to the plain, and there made their camp. In the same spot was founded the city. It was quite as good a locality as any other, no doubt they thought, considering that the whole region alike was nothing but a plain of sagebrush. Indeed, you cannot see the lake at all from the city, except by going up upon the "bench" of higher ground to the northward and eastward, whence it appears only as a line of distinct color between the dusty olive of the wide foreground, and the vague blue of the distant hills.

The habitations of the pioneers were not built hastily and at random. Brigham Young caused a town-site to be carefully surveyed and accurately laid out, and it was done on a generous scale. The streets were made one hundred and thirty feet wide, placed true to the points of the compass and crossing one another at right angles. Each square contains ten acres, so that when the Madame and I merely walked "around the block" while I smoked a post-prandial cigarette, we tramped precisely half a mile. A square of nine blocks was made to constitute a "ward" —now the city has twenty-four—presided over by a bishop of the church. Despite his title, he was more a temporal than a spiritual head, deciding all small matters in dispute in those simple first days when there was no appeal, nor desire for one, from ecclesiastical decisions to civil judgment. Even yet, this ward classification enters largely into the social constitution of the city.

When the streets and wards had been determined each pioneer was given an acre and a quarter as a town lot, and as much outside land as he could occupy. This accounts in a great measure for the ample space and farm-like appearance of the grounds around most of the houses in this widely dispersed city.

To make this real estate of value, however, water for irrigation must be brought to it. This was supplied by the "City" creek flowing down from Emigration cañon, whose current was led into ditches all over the new colony, and still fills the roadside gutters with sparkling streams, nourishing many gardens, and the roots of the long lines of varied shade trees, whose boughs almost reach over the thoroughfares.

All the brethren worked in common at this ditching, and it was done so soon that within a few days after their arrival seeds had been put in the ground for the first crop.

"Yes," says the Madame, as I relate this history, "and they say that old Jim Bridger watched them cynically and said they were a pack of—well, no matter what kind of fools, and that he would give a thousand dollars for the first ear of corn raised there."

"That's said to be true," I assent.

"But he had to acknowledge the corn," Chum puts in—and flees!

Formerly this water alone was available for domestic purposes and drinking, as well as for irrigation, and even yet the poorer part of the population dip it up at the curbstone for daily service. But the introduction of pipes and hydrants has now superseded this old way, though the water is no better; for table use, therefore, the sweet pure beverage drawn from very deep wells is preferred. Experiments are making in this respect to artesian wells also.

The houses built by the first settlers were mainly log cabins, and some relics are still to be found hidden away in blossoming orchards. The Spanish-American *adobe* house was also a favorite, and has continued so to the present, though instead of almost shapeless chunks of mud, plastered in Mexican fashion, regular unburnt bricks are made by machinery. These *adobes* are twice the size of ordinary bricks, and the wall into which they are formed is made twice as thick as one of burned bricks would be. Of course this material lends itself to any style of architecture, and many of the elaborate buildings, as well as cheap cottages, are made of it, the soft gray tint of the natural *adobe*, or the gentle tone of some overlying stucco harmonizing most tastefully with the crowding greenery. Low houses, with abundant piazzas and many nondescript additions, are the most common type in the older part of the town; and over these so many vines are trained, and so much foliage clusters, that one can hardly say of what material the structure itself is formed. The residences recently built have a more eastern and conventional aspect, and some are very imposing; but, big or little, old or new, it is rare to find a house not ensconced in trees and shrubs and climbing plants, while smooth, rich, well-shaven lawns greet the eye everywhere in town, in brilliant contrast to the bleak hills towering overhead just without the city. As for flowers, no town east or west cultivates them more universally and assiduously.

"There are no florists here," says the Madame.

"And no need for any—each man has his own plants if not the luxury of a greenhouse."

Salt Lake City, then, is beautiful—a paradise in comparison with the buffalo plains or the stony gulches in which the majority of Rocky Mountain towns must needs be set. Nor is there any question as to the fact that this is wholly to the credit of the Mormons—not because they were Mormons, but because they were diligent and foresighted, and came hither not to make a fortune and escape, but to stay and build up pleasant homes and a prosperous commonwealth. Any other set of men might have done the same; but certainly no other set of men *did*, for to no others was presented the same compelling motive.

The suburbs—except toward the rocky uplands northward—grade off quite imperceptibly, the streets continuing straight out into country roads between dense jungles of sunflowers,—glorious walls of gold, and green; and in these suburbs you may find some of the queerest, most idyllic cottages.

The two broad distinctions of "Mormon" and "Gentile," are not enough to represent the elements of Salt Lake society. At least three divisions ought to be counted. First, the Latter-day Saints; second, the seceders from the Mormon Church; third, the Gentiles — respectable people, mostly attendants at Christian churches.

"Such a classification must make queer comrades," remarks Chum, as we sit talking over these matters.

"I should say so," I reply, "the Jew becomes a Gentile, and the Roman Catholic becomes a Protestant, making common cause with Calvinism against the hierarchy of the Temple."

"I do not suppose," the Madame observes, "that they can sink their own little differences, although allied in one fight; so that society must necessarily be divided into a lot of little groups, and thus lose a great deal."

"Yes, the people who profess no religious adherence have rather the easiest time of it in Salt Lake, I believe."

The non-Mormon part of the citizens probably enjoy themselves more than they would if the isolation of the locality did not compel them to be self-centered and contrive their own amusements to a great extent. It is a society made up of the families of successful merchants and mining men, of clergymen and teachers, of the officers of the army stationed at Camp Douglass, and the representatives of the government in the judicial and other territorial offices. This composition, it will be seen, presupposes considerable intelligence and cultivation. It was not until Gentile gold came in to break up the old custom of barter, that the resources of the Mormon community became really available either to themselves or to others.

Utah has always been pre-eminently an agricultural district. Out of her one hundred and fifty thousand people probably one hundred and twenty thousand are now farming or stock-raising in some capacity or other. When you look down the valley from the city, your eye takes in a wide view of fields, orchards and meadows, green with the most luxuriant growth, and marked off by rows of stately trees or patches of young woodland. All these farms are small holdings, and though cultivated

GREAT SALT LAKE.

by no means scientifically, have long produced well up to their several capacities.

The exports of all sorts of grain, produce and fruit are large, and increasing, thanks to this new railway of ours and its encouraging rates of freight.

The Mormon leaders, and particularly Brigham Young, at first opposed any attempt at a development of the mineral resources of the territory, though the latter is said to have been well informed of their character and value. He forbade all mining to his people, and would have closed the mountains to Gentile prospectors if he had been able. So far as a desire existed to avoid the evils of a placer-working excitement, drawing hither a horde of gold-seekers, this course was a wise one; but as years went on, it was seen by the shrewder heads among the Mormons themselves that this abstinence from mining was harmful. There was no cash in the treasury, and none to be got (I am speaking of early days). If a surplus of grain was raised, or more of any sort of goods manufactured than could be used at home, there was no sale for them, since at that time, the market was so far away that the profits would all be lost in the expense of transportation.

It is funny to hear the tales of those days. Business was almost wholly by barter, and payments for everything had to be made by exchange. A man who took his family to the theatre wheeled his admission fee with him in the shape of a barrel or two of potatoes, and a young man would go to a dance with his girl on one arm and a bunch of turnips on the other with which to buy his ticket. Gentile emigrants and settlers soon began to bring in coin, but the relief was gradual and inadequate.

Finally, about fifteen years ago, it was publicly argued by more liberal minds that the only things Utah had which she could send out against competition were gold and silver. When, from preaching they began to practice, and enterprising men encouraged outside capital to join them in developing silver ledges in the Wasatch and Oquirrh ranges, then Salt Lake City began to rouse herself. Potatoes and carrots and adobes disappeared as currency, and coin and greenbacks enlivened trade which more and more conformed to the ordinary methods of American commerce.

One quite legitimate means taken for centralizing of trade was the establishment, twenty-five years ago, of Zion's Coöperative Mercantile Institution. In the early days it was extremely difficult for country shopkeepers to maintain supplies when everything had to be hauled by teams from the Missouri river, and the most extortionate prices would be demanded for staples, whenever, as frequently happened, a petty dealer would get a "corner" on some article. A few great fortunes were quickly made, but a stop was put to this by setting on foot the coöperative establishment, which was imitated in a small way in many rural settlements.

The design of this institution was to furnish goods of every sort known to merchants out of one central depot in Salt Lake City under control of the Church and partly owned by it. This was a joint-stock "coöperative" affair, however, and the capital was nearly a million dollars. The people were advised from the pulpit to trade there, but they would have done so anyhow, for the "Coöp," as they called it, was able to reduce and equalize prices very greatly. Branches were established in Ogden, Logan, Soda Springs, and lately a warehouse built in Provo. These and other additions were rapid. The central salesrooms in this city now occupy a four-story brick building, three hundred and eighteen feet long by ninety-seven wide, where every species of merchandise is to be found. In other quarters are a drug-store, a shoe factory (supplied by its own tanneries and running one hundred and twenty-five machines propelled by steam), and a factory for making canvas "overall" clothing. Altogether about two hundred and fifty persons are employed, working reasonable hours and for reasonable wages. The stock, which originally was widely scattered, has been concentrated for the most part in the hands of a few astute men, who are credited with large profits. There is an air of great prosperity about the institution, whose business is stated to reach five million dollars annually, derived almost wholly from Utah.

Though this concern had a practical monopoly at first, as soon as the railways came to Salt Lake, individual merchants could sell goods about as cheap, and opposition to it arose.

Religious competition has arisen. Among the first of these local Protestants was a mission of the Roman Catholics. Now they have a considerable colony here and in Ogden. The St. Mary's Academy, in charge of the Sisters of the Holy Cross, has a large building, beautiful grounds, and the reputation of being a first-class higher school for girls. There is a school for little boys in the same enclosure. The boarders at the Academy amount to about one hundred annually, and the day scholars to one hundred and fifty. The Sisters of the Holy Cross also have charge of a large and finely-conducted hospital in the eastern part of the city.

Another hospital is the St. Marks, supported partly by monthly dues from miners, and otherwise by special contributions. This is in charge of the Episcopal Church, which has been active in Utah for many years under the guidance of Bishop Tuttle. St. Mark's School, belonging to the local church organization, had three hundred and thirty pupils during its last term. The Methodist Episcopal denomination, also, has churches scattered about the territory and schools in Salt Lake City, among the rest night schools for Chinamen, who are an important element of the population. The Presbyterian Church has set up here a

Collegiate Institute, owning property worth about seventy-five thousand dollars and giving instruction to about two hundred pupils, from the primary to a high-school grade. This is unsectarian, as, I suppose, are all the rest so far as any active religious pressure is brought to bear. The most exclusive school, probably, is that sustained by the Hebrew Society. As in other western towns the Jews are in large force in Salt Lake City, their characteristic names occurring on many a signboard.

The Mormons themselves sustain a system of public schools, in which, in addition to the usual branches, the tenets of their faith are taught. These schools are well conducted and will compare favorably with those in any city the same size.

Salt Lake City is a great center of wholesale trade in provisions and textile fabrics not only, but in machinery and mining supplies. She has smelters; a lead-paint factory; foundries and boiler works; sampling-mills handling two hundred tons of ore a day, brought from far and near; breweries, carriage and furniture shops; and all sorts of small factories. Traction engines and locomotives, if not wholly built there, are reconstructed; and complicated machinery of other sorts is manufactured. Her salt business, now that a liberal minded railway has come to her relief, is likely to become of the greatest importance, which will be a benefit to her, not only, but to all the smelters and chlorodization works in the Rocky Mountain region.

The city grows rapidly and becomes daily more cultivated and beautiful, and less *outre*. Every appliance of civilization is utilized, and she has the best hotels by far between Denver and San Francisco—some think even better than either, but that is an extravagant estimate. Statistics show that six hundred new houses were built, five hundred and seventy-four of them dwellings, at a cost of $1,636,500. By the time the next census is taken, in 1890, she may contain fifty thousand inhabitants. The Madame and I thought we would rather make our home in Salt Lake than in any town west of the Plains; but Chum cast his vote in favor of Denver.

XXXVI

SALT LAKE AND THE WASATCH.

> Behind, the silent snows; and **wide below**,
> The rounded hills made level, lessening down
> To where a river washed with sluggish flow
> A many-templed town.
>
> —BAYARD TAYLOR.

ONE day we all went out to the great Salt Lake, as in duty bound. You might as well go to Mecca and fail to see the tomb of the Prophet, as to visit Deseret and avoid the lake. It is a ride of twenty miles by rail, and the fare for the round trip is only fifty cents. Two trains are run every day in summer, and they are especially well-filled on Sundays. The cars used are chiefly open ones, with seats crosswise, like those run to Brighton and the other Beaches from New York, and it would be good fun in itself to go racing in this free way across the breezy desert between the city and the lake, even if there were not the salt waves at the end of the journey.

For, of course, the only object in going to the lake—or at any rate the prime object—is the bathing. There are two or three landings, all much alike, and not far apart; which one it was we stopped at, I have forgotten, and it doesn't matter. One is called Garfield and another Black Rock, after a great cubic mass of lava that stands out of the water a little way from shore like the end of a huge ruined pier.

Unfortunately it is impossible to make trees grow at the shore. The water and the soil are too bitterly salt; moreover, there is no fresh water in the rocky hills of the Oquirrh that tower straight up from the beach, and irrigation is thus forestalled. In lieu of this, a few wide-verandahed houses and open sheds exist, with several booths made of boughs and evergreens, under which are long tables and benches for the accommodation of those who bring their lunches. Nearly every day you will see these bowers half-filled with picnic parties who have come to spend the day; and there are frequent excursions from the city, where large parties go out in the evening, dance all night and return by a special train in the early morning.

At the edge of the water are rows of dressing closets where the bathing suits are donned and whence you go by stairways directly into the water. No special hours are thought preferable. Men and women go in under a noonday blaze that makes the brain swim on shore, and assert that their bare heads suffer no discomfort. We thought their crania must be harder than ours, however, and postponed our dip till the cool of the evening.

While the danger of sunstroke seems very small—the rarity and purity of the air get the credit for this—the lake is a treacherous place for swimmers. The great density of its waters sustains you so that you float easily, but for the same reason swimming ahead is very tiresome work. Moreover, fatal consequences are likely to ensue if any considerable quantity of the brine is swallowed. It not only chokes, but is described as fairly burning the tissues of the throat and lungs, producing death almost as surely as the inhalation of flame. Of course this occurs in exceptional cases only, but many persons suffer extremely from a single accidental swallow. I remind the Madame of this as I lead her rather timid feet down the steps, and add that most of the sufferers hitherto have been women.

"That's because they can't keep their mouths shut even on pain of death," remarks Chum, with malice aforethought. For this remark, some day, I have no doubt, he will be called to account, by my wife, who seems more worried at present, however, to keep the brine out of her hair than out of her mouth.

The powerful effect of this water is not surprising when one remembers that the proportion of saline matter—about twenty per cent,—in it is six times as great as the percentage of the ocean, and almost equal to that of the Dead Sea, though Lake Oroomiah, in Persia, is reputed to contain water of a third greater density yet. This density is due mainly to common salt held in solution, but there are various other ingredients. In Great Salt Lake, for example, only 0.52 per cent. of magnesia exists, the Dead Sea having 7.82 per cent.; of lime, Salt Lake holds 1.80 per cent., while the Dead Sea contains only a third as much. As you look into it the water seems marvelously transparent, so that the ripple-marked sand and pebbles at the bottom show with strange distinctness. This is usually adduced as an evidence of its purity, and in one sense it is so; but it is also the result of its density, since the invisible particles of salt in it, catch and carry the light to far greater depths than it would be able to penetrate in distilled water, which, also, would be perfectly clear. The crystal clearness and intense color of the water of the Mediterranean is noticed by all travellers; but it is also the fact that the Mediterranean is considerably salter than the open Atlantic.

Great flocks of gulls and pelicans inhabit the upper part of the lake and breed upon the shores and islands; what they all find to eat is a mystery. No vegetation can survive where the spray of these bitter waves has dashed, save a miserable little saltwort and a melancholy species of *Artemisia*, whose straggling and thorny limbs appear black and burnt on the scorching sands. Salt is made in great quantities in summer, by the simple process of damming small bays and letting the enclosed water evaporate, leaving a crust of crystallized salt behind. Several thousands of tons are exported annually, and great quantities used at home in chlorodizing silver ores.

I think few persons realize how wonderfully, strangely beautiful this inland, saline sea is. Under the sunlight its wide surface gives the eye such a mass of brilliant color as is rarely seen in the temperate zone. Over against the horizon it is almost black, then ultra-marine, then glowing Prussian blue; here, close at hand, variegated with patches of verdigris green and the soft, skyey tone of the turquoise. If the lake were in a plain (remembering the total absence of forest or greensward) doubtless this richness of color would not suffice to produce the effect of beauty, but on every side stand lofty mountains. They seem to rise from the very margin to their riven, bare and pinnacle-studded crests spotted with snow, though some of them are miles beyond the water's edge.

Two mountainous islands stand prominently in view at the lower end of the lake—Church and Antelope. On the former some two thousand head of cattle are pastured. The latter has a less prosaic history, though at present similarly utilized as grazing-land. When the Mormons first came hither they wintered their cattle and horses upon it. The eastern side of the island contains some farming land, and a quarry of roofing slate.

An obliging gentleman told us all about the island, and also gave an account of what must have been an exciting chase. He said that until two or three years ago there roamed upon the island a remnant of the horse-herds once pastured there, numbering fifty or sixty horses and mares. These were as wild as wild could be, and grazed upon the western side of the island, which is very broken and rocky, and traversed by narrow trails that the horses had worn in the hillsides. It was decided to attempt to capture some or all of these horses and a novel method of snaring was adopted. Nooses were made at the ends of long lines which were securely anchored; the nooses were then hung in the bushes in such a way as to overhang the trail at the proper height. Several mounted men then got behind a few of the wild herd, and drove them as furiously as they could frighten them forward along the narrow trails. Overcome with terror the leading animal never saw the dangling rope, but rushed his head through the noose and was instantly jerked off the trail. Tearing wildly past him half a dozen others, one by one went into as many consecutive snares and were caught.

As each horse was caught, one of the pursuers would hasten to him as rapidly as possible, fasten the end of the lariat to the horn of his saddle, and then lose no time in loosening the noose about the captive's neck, which by that time would have choked the poor beast almost into insensibility. This done, he would leave the wild and tame animals tied together, to fight it out, and hurry on to help his companions. In this way several horses were captured, and proved very docile and capable when put in the harness.

The story has scarcely been concluded, when we are called to our homeward-bound train. It is just at sunset—the western horizon a fountain of fiery gold seen through a saffron veil of ineffable splendor. The air seems to become saturated—thick with color throughout the whole space between us and the horizon. The mountains shine through this veil in a sharply defined mass, not a single feature visible, but their whole silhouette washed in with a flat tint of marvelous softness and inimitable delicacy. Yet it changes, almost every instant, and gradually, as the orb disappears behind the island, and the cloth of gold laid down for his feet across the lake, is drawn away, the island-hills and the jagged sierras beyond settle into cold ashy blue, and the coolness of approaching night already fans our cheeks.

Another day we made an excursion up into the cañon of the Little Cottonwood to Alta—a mining town known all round the world. The place is not only entertaining in itself, but in its neighborhood are a large number of easily accessible gorges, lakes and hilltops full of artistic material and of trout fishing; or, if the tourist goes late in the season, of good shooting and ample opportunity for dangerous adventures in mountaineering. The Little Cottonwood is one of those great crevices between the peaks of the Wasatch range plainly visible from Salt Lake City, and distinguished by its white walls, which when wet with the morning dews gleam like monstrous mirrors as the sunlight reaches them from over the top of the range.

We took the early morning train down to Bingham Junction, so called because branch roads diverge here, not only to Alta, our destination, but also to Bingham, a mining camp opposite, in the Oquirrh, which has attracted much attention in the past and still has very profitable mines, with many peculiarities of great interest to the specialist. Here at the Junction stood awaiting us a locomotive heading a train made up of almost every kind of car known to rolling stock. Whisked away past fields of lucerne we were quickly climbing the foothill benches and entering the mouth of the cañon, where the train came to a standstill underneath an ore-shed and alongside of a beer-saloon. In front of the saloon stood on slender rails two or three of the queerest vehicles it was ever my fortune to ride in. If you can imagine the body of a three-seated sleigh, with its curled up splash-board, mounted upon a hand-car and rigged with a miniature "boot" behind, you will have an idea of these vehicles in which we were to finish our trip up the eight miles of cañon remaining. The motive power consisted of two black mules, harnessed tandem, and the driver was the conductor of the train, who disguised himself so effectually in a big hat and bigger duster that it was a long time before we discovered his identity.

The walls of this cañon are extremely lofty, and in places almost vertical. Though in crevices and ledges here and there some fearless bushes and trees have maintained a foothold, yet there are large spaces of almost upright slope, wholly bare of the least soil or vegetation, and smoothed by the waters that drip over them, the sliding avalanches that sweep their faces, and the fierce winds that polish them under streams of sharp-grained dust. Whiter precipices I have never seen, and the rock lies in long layers, that in the case of sedimentary rocks we would call strata, inclined at a very steep angle against the higher heart of the range within. Here, too, are the usual lines of cross-cleavage, and in these lines, as well as between the layers, water finds itself able to penetrate more or less easily. Hence the frost during past ages has slowly cracked off great masses of exposed cliff and hurled them down. This rock does not crumble, as would the lavas, but falls in masses, and with these the bottom of the cañon has been gradually filled up. The water of the creek finding its way over and among the great pieces, never ceases to be a cataract, or has a moment rest from its foaming haste; and our tramway squirmed and dodged among angular fragments, each as big as a house, which had fallen so recently as yet to be lying on top of the ground.

It is by splitting to pieces these great detached droppings of the cliff—solid fragments of the original granite cliff,—that the contractors get the fine building stone of the Mormon temple in the city. There is no need to open any quarries. It is only necessary to drill and blast these big stones lying on the surface, and the demands of a hundred temples would not exhaust the supply. Men were at work as we passed, splitting out blocks that were dragged by stoneboats, or sent along the tramway down to where they could be loaded upon the railroad cars. Until three years ago every bit of this stone was hauled all the way to Salt Lake City by bullock teams, and the great expense and labor account both for the large expense and the slow progress of the mighty structure.

A mile or so above Wasatch station, the tramway entered a snow-shed; and with momentary exceptions, it never got out of it for seven miles. To the sight-seer this was discouraging; but it was compensated by the coolness, for in the stillness of the cañon, the sunshine, reflected from the dazzling walls, was fiercely hot, and our occasional emergencies into it was like passing before the door of a blast furnace. These sheds are said to have cost one hundred thousand dollars, though the timber was close at hand and sawed in the cañon. They were necessary, for this is a gorge famous for its depth of snowfall and its avalanches. It required two hours to toil through the sheds and at the end we found as peculiar a scene of human life as could well be imagined. The cañon "bends" here, in an almost complete circle of heights, some of which reach, stark and splintered, far above timber-line. At the disbandment of

General **Connor's** regiment of Californian troops in 1863, they scattered through the mountains and among other places came here. Prospecting the higher slopes, silver ore was discovered, and a host of miners came in, and began digging on all the hills. The famous "Emma," the "Flag Staff," and dozens of other mines were opened. A town, well-called *Alta* (high), sprang up, and filled all the level land at the head of the valley, while buildings, and machinery and dumps dotted the mountain sides to their topmost ridges. Long paths had marked the ruin of avalanches before this, but when, to supply timber for the mines and the cabins, the mountain sides were denuded of their forests, large areas of deep snow became loosened in every great storm, and slid with crushing force, tearing up and carrying everything before it, to the bottom of the slope. Once the whole corner of the town was swept clean away; again and again miners lost their buildings at their tunnel entrances. Little work could be done in winter yet many stayed in Alta, isolated from the world, and at the mines, and many and many a one lost his life, to have his body found in a horrible condition when the winter was over. Then in the spring, when the frost was loosening the ground, and the melting snow was pouring a thousand waterfalls down the sides of the cañon, the snowslides were succeeded by the giving away of masses of soil and loose rocks, which came headlong into the bottom of the cañon. One such avalanche of rocks was pointed out to us which had slid down the opposite mountain with such force as to carry it clear across, and almost a hundred feet up the hither slope, sweeping away the tramway, sheds and all.

Meanwhile the original owners of the mines had sold them in the most prominent cases, for enough to make the men wealthy. Companies had been formed, the stock had been put upon the market, and the usual history of a mining camp was gone through. The "Emma," in the hands of a company of English capitalists, was made notorious by litigation, and for a long time was shut down. Now, however, a new era is beginning. Work has been resumed on many lodes that for years have been idle, and arctic Alta may yet range herself among the foremost silver-producing localities of the territory. We were all glad we went up there, yet were quite ready at four o'clock to return.

When we took our seats in the little sleigh-like car, no mules stood sedately tandem in front of it; and before we understood that we were ready, behold we were off! It was merely the loosening of a brake, and the car began to roll swiftly down the track. That was an exhilarating ride! Whisking round the curves, rattling through long tunnels, dodging out into the sunlight to catch a glimpse of a sparkling waterfall, or a bit of plain seen away down the cañon, then back again into the tunnel, where gophers and chip-munks and cotton-tails were continually perking up their heads and then settling into some small cave of refuge as we rushed past—on and on, down and down in the face of the stiff breeze and under lofty walls, without an instant's check, until we glided into the little terminus, just twenty-five minutes out of Alta!

But our gravity railroading was not done yet. A small passenger car stood at the head of the railway track by which we had come up from the valley. As soon as we had entered it, our jolly driver-conductor (there was no gravity about him!) loosened the brake and we rushed off again like the ghost of a train, without engine or engineer, and went spinning down the tortuous track for a dozen miles to Bingham Junction. It was just as good fun as coasting—and better, for you did n't have to drag your own sled back up hill again.

XXXVII

AU REVOIR.

> End things must, end howsoever things may.
> —BROWNING.

THIS was our last excursion, and all three of us knew it as we gathered in our own coach again at Bingham Junction.

"At last," remarks Madame, cheerfully—she is thinking that before many more days an apple-cheeked little damsel in far New England will be back in her arms—"we have come, sir, to the final chapter. The emptiness of your utmost corner-pigeon-hole will reproach you no longer. A few days more and *Finis* will be written across the completed manuscript, and our glorious cruise will be a thing of the past. Meanwhile, sir, remember your 'Cocheluuk,'—

> 'Act, act in the living present,
> Heart within and God o'er head.''

"For instance?" I ask, after this homily.

"Observe, and make a note of, these great meadows of rich grass and the russet areas where hay has been cut. Note how, among the plumey masses left standing scarlet flowers are burning like coals—I wonder if prairie fires ever originate from their igniting the dry and feathery stalks! See how the Jordan flows stately down the center of this wide mountain trough, its banks crowded with farmhouses, each in its little copse of trees. Long lines of Lombardy poplars mark the boundaries of many farms and willows show where the big irrigating ditches pass or rivulets trickle. All these things are of the highest interest, and imply a mass of statistics you ought busily to gather and carefully to record in tables of precise and copious information."

"Eh?" I say.

What is the matter with the Madame? Is she making fun of somebody whom she ought to hold in a respect almost amounting to awe? Feeling that I ought to assert myself I gently hint that this is my affair, and her help is uncalled for in the matter of book-making; that her own department is wide enough for all her energies; and that—

But here Chum interrupts in that strix-like way of his which always so commands attention that one must listen whether or no.

This young man is possessed of a family heirloom in the shape of several hundred traditions of a more or less mythical grandfather. Some of these tales are distinctly poetical, while none of them are prosy. It is one of the traditions coined in the ingenious brain of this talented old gentleman with which we are now regaled, *apropos* of the matter in hand.

The old gentleman, it appears, was once—but let his heir-apparent—

"Who," the Madame interrupts maliciously, "has very little hair apparent."

"Let him," I say, ignoring the insinuation, "tell his own story."

"Why it was this way, as you very justly remark. The old gentleman was once captured by the Indians, who, instead of scalping him, decided to make him a beast of burden. They, therefore, loaded him down with cooking utensils, the most prominent article of which was the useful, but heavy frying-pan known in the vernacular as 'skillet.' Each Indian deposited upon my grandfather's venerable and enduring back, his skillet. The old gentleman dare not protest, but meekly submitted and trudged off under his Atlas-like burden. After two hours hard marching, however, he resolved to argue the question, so he shouted imperatively,

"'Halt!'

"The Indians paused in wonder. The venerable victim climbed upon a fallen tree and delivered his famous forensic effort, as follows:

"'Mr. Injuns! I have a proposition to make. *I move that every Injun carry his own skillet.*'

"The modesty and yet fairness of this proposition met with an enthusiastic reception and every Indian after that 'carried his own skillet,' which commendable example it would be well for all to follow."

"That's a good story!" I remarked. "A good moral story! This expedition, my dear Madame, was for fun, not for geographical pedantry; and my book shall make no pretense to be a cyclopædia, a guide, or a useful companion of any sort, but just a jolly story of a care-forgetting vacation. If it jogs the curiosity, whets the appetite, nerves the fingers, weak through long toil in tying, and untying purse strings, to come and see what *we* have seen, that is all the effect that can be expected; and this much done, the traveller who follows our uncertain trail will find out far more for himself than we ever could hope to tell him. Seeing Colorado, no matter how briefly,

'Of her bright face, one glance will trace
A picture on the brain,
And of her voice in echoing hearts
A sound must long remain.'

"But here we are at Salt Lake, and home again, for one more gay dinner in the red-walled car; one more gay evening under the cool stars; one more night's rest in the queer little stateroom. To-morrow, Chum, old 'friend and fellow-student,' in lonely grandeur you will be taking the long-to-be-remembered 'special' swiftly back to Denver, while the Madame and I are rolling away to the Golden Gate. Fill your glasses. And what shall the toast be? **The** God-wrought landscape we **have seen? The wide-awake people we have known?** The **splendid railroad whose** achievements we know and **of whose** hospitality we have partaken? The glorious 'good times' we've had? The stores of health we have **laid away?** Ay, all these and more. **Let us toast** each other; and then—

Good Night!